536.7015 Smi 199167
Smith.
Elementary statistical
thermodynamics.

The Lorette Wilmot Library
Nazareth College of Rochester

ELEMENTARY STATISTICAL THERMODYNAMICS

A Problems Approach

ELEMENTARY STATISTICAL THERMODYNAMICS

A Problems Approach

NORMAN O. SMITH

Fordham University
Bronx, New York

LORETTE WILMOT LIBRARY
NAZARETH COLLEGE

PLENUM PRESS · NEW YORK AND LONDON

Library of Congress Cataloging in Publication Data

Smith, Norman Obed,
Elementary statistical thermodynamics.

1. Statistical thermodynamics. I. Title.
QC311.5.S584 1982 536′7′015195 82-18131
ISBN 0-306-41205-5
ISBN 0-306-41216-0 (pbk.)

199167

©1982 Plenum Press, New York
A Division of Plenum Publishing Corporation
233 Spring Street, New York, N.Y. 10013

All rights reserved

No part of this book may be reproduced, stored in a retrieval system, or transmitted in any form or by any means, electronic, mechanical, photocopying, microfilming, recording, or otherwise, without written permission from the Publisher

Printed in the United States of America

536.7015 Smi

PREFACE

This book is a sequel to my *Chemical Thermodynamics: A Problems Approach* published in 1967, which concerned classical thermodynamics almost exclusively. Most books on statistical thermodynamics now available are written either for the superior general chemistry student or for the specialist. The author has felt the need for a text which would bring the intermediate reader to the point where he could not only appreciate the roots of the subject but also have some facility in calculating thermodynamic quantities. Although statistical thermodynamics comprises an essential part of the college training of a chemist, its treatment in general physical chemistry texts is, of necessity, compressed to the point where the less competent student is unable to appreciate or comprehend its logic and beauty, and is reduced to memorizing a series of formulas. It has been my aim to fill this need by writing a logical account of the foundations and applications of the subject at a level which can be grasped by an undergraduate who has had some exposure to calculus and to the basic concepts of classical thermodynamics. It can serve as a text or supplementary reading for a course, or provide the means whereby one could become conversant with the subject on his own, without the benefit of an instructor. To this end the text is profusely illustrated with worked examples, and each chapter concluded with numerous problems, all of which have the answers provided at the end of the book and many of which have also the details of how they were obtained. It is my firm belief that a student cannot be considered to understand a quantitative subject such as thermodynamics until he can work numerical problems in it.

This is an elementary treatment and does not include any reference to ensembles. As a consequence only atomic crystals and ideal gases can be said to have been fairly well covered. This, to some, will be a serious limitation, but at the undergraduate level, and perhaps at the beginning graduate level, it was considered acceptable. Anyone with a working knowledge of the topics covered in this book should have an adequate foundation on which to study ensembles on a later occasion.

After developing the subject of probability from first principles, the elementary statistical mechanics of systems of distinguishable particles is considered, the Boltzmann distribution law derived, and the statistical basis of entropy treated. The statistical treatment is then extended to systems of nonlocalized particles, which leads into the subject of ideal gases. The calculation of the partition functions for the various kinds of energy is then developed at length in order to show how thermodynamic functions are computed. This includes a section on the determination of the center of mass and moments of inertia of polyatomic molecules in more detail than is usually found in textbooks for chemists—in the belief that such calculations are instructive and well within the reach of the average student. The ready accessibility of desk calculators now takes the tedium out of computations of this sort. It is in the calculation of the thermodynamic functions where the student who has made the effort to understand the framework of the subject is suddenly rewarded. Finally the subject of chemical equilibrium is considered, with somewhat more emphasis on the influence of symmetry than will generally be found elsewhere.

The choice of units is a vexing one for any author of a scientific book. If he wishes to be fashionable he will use the SI units and risk the ire of many practicing chemists who see the clumsiness which sometimes comes with it, if followed to the letter. On the other hand, if he ignores it he is out of step with current international trends. I have chosen to use the SI units except when I have felt it to be too awkward: the atmosphere, for example, has been retained as the pressure unit, and molecular weights quoted in grams per mole.

It is a pleasure to acknowledge that I have benefited by reading the various approaches to the subject in the books by Professors Henry A. Bent of North Carolina State University, Malcolm Dole of Northwestern University, Leonard K. Nash of Harvard University, and, especially, Maurice H. Everdell of the University of Aston in Birmingham, whose recent book is a gem. Finally I express my appreciation to Fordham University for a Faculty Fellowship with which to undertake the task.

N. O. S.

CONTENTS

INTRODUCTION

Thermodynamics is the study of heat and its relation to other forms of energy. There are generally two approaches to the subject, which we shall refer to as the classical and statistical. Some prefer that they be studied separately in order not to confuse the ideas, others prefer that they be studied together, or at least in parallel, in order to see how one augments the other. Both approaches are based on experimental data, but the kinds of data used are different. Classical thermodynamics uses data based on the properties of matter in the bulk—such as density, heat capacity, vapor pressure—whereas statistical thermodynamics uses properties of individual molecules such as bond length, vibration frequency, and symmetry. When, for example, classical thermodynamics tells us that the entropy of formation of CO_2(g) is 2.87 J/K mol at 298 K and 1 atm, this information has come from measurements with bulk quantities of graphite, O_2(g) and CO_2(g), with no enquiry into what there is about these three substances which gives the observed value. Statistical thermodynamics, on the other hand, determines what there is about the individual atoms and molecules which leads to the measured result. Purely classical thermodynamics ignores molecular complexity and quantum mechanics (which it pre-dated); statistical thermodynamics, since it is concerned with the individual molecules, leans heavily on the results of quantum mechanics. Further, since matter is composed of very large numbers of molecules, and since these molecules, even in pure substances, differ from each other in some respects, it is necessary to average their behavior in order to predict their bulk properties—

hence the adjective "statistical." It is a rewarding experience to see how these two very different avenues lead to the same results.

Chapter One

STATISTICAL MECHANICS OF DISTINGUISHABLE PARTICLES

1.1. PROBABILITY

In order to develop the tools for evaluating thermodynamic quantities by statistical methods we begin with the observation that we are dealing with immense numbers of particles (atoms or molecules) of matter. Each particle has available to it an immense number of possible energies of various kinds, but, left to itself, an isolated collection of such particles will eventually come to an equilibrium state in which the number of particles in any one state will thereafter remain constant. If the collection or system is isolated, classical thermodynamics tells us that it will, quickly or slowly, come to that state with the greatest possible entropy. This is, in fact, the famous second law of thermodynamics, often written $dS_{E,V} \geq 0$, where S, E, and V are entropy, energy, and volume, respectively. At the same time we know from everyday experience that any spontaneous change involves a progression from a less to a more probable state. The conclusion is inevitable that there ought to be some connection between entropy and probability. We are thus led to examine the subject of probability and, in turn, enquire how one can determine which of a great many possible states is the most probable one. We can then anticipate that a knowledge of the most probable state will lead us to its entropy and to other thermodynamic properties.

We begin with an elementary exercise, the tossing of a pair of dice, each of which is a cube, the faces of each cube being marked 1, 2, 3, 4, 5, and 6. Assuming that the dice are not loaded, the chance of any one of the six faces lying face up when the dice are thrown is the same as that of any other. One "throws a two" when both dice show 1, and this can happen in only one way. One "throws a three" when a 2 and a 1, or when a 1 and a 2 show, that is, in two ways. One "throws a four" when the combinations [1,3] or [2,2] or [3,1] show, that is, in three ways. All the possibilities are as follows:

"Throw"	2	3	4	5	6	7	8	9	10	11	12
Number of ways	1	2	3	4	5	6	5	4	3	2	1

The total number of ways is 36. Thus, of this total, six ways result in throwing a seven. The probability of throwing a seven is thus 6 out of 36 or 1/6—the ratio of the number of favorable events to the total number of possible events. The most probable event, namely, throwing a seven, is the one which can be accomplished in the greatest number of ways. Letting W stand for each of the entries under "Number of ways" we can say that $W = 6$ for throwing a seven, and that the total W, or $W_{tot} = 36$.

EXAMPLE 1.1. Suppose that the pair of dice referred to above were regular octahedra instead of cubes, with faces numbered 1 to 8. Find (a) W_{tot}, (b) W/W_{tot} for the most probable throw.

ANSWER. (a) Each die can land in eight ways, so $W_{tot} = 8^2 = 64$; (b) Throws of two to sixteen are all possible, with W progressing from 1 for a throw of 2 to 8 for a throw of 9, then diminishing to 1 for a throw of 16. The most probable throw is 9, for which $W/W_{tot} = 8/64$ or 1/8.

Consider now the question of the number of ways a given number of balls can be distributed in a given number of boxes.

Let us use three balls and four boxes. We shall suppose that the balls are distinguishable—for example, differing in color—and so label them a, b, and c, (*a*mber, *b*lue, *c*rimson). The boxes will be referred to as numbers 1, 2, 3, and 4. Now a can be placed in any of four boxes. For each of these, b can also be placed in any of the four boxes, so there are $4 \times 4 = 16$ ways of placing a and b. For each of these there are 4 ways of placing c, so the total number of ways is $4^3 = 64$. Suppose, however, that we impose the restriction that there shall be two balls in box 1, one in box 2, and none in boxes 3 and 4. This is a serious limitation which reduces the number of ways to three, namely, a and b in box 1 and c in box 2, a and c in box 1 and b in box 2, and b and c in box 1 and a in box 2. In general, when a total of N distinguishable objects are in boxes 1, 2, 3, . . . , i, respectively, combinatorial mathematics tells us that the number of ways W is given by

$$W = \frac{N!}{n_1!n_2!n_3! \cdots n_i!}$$

which is abbreviated

$$W = \frac{N!}{\prod_i n_i!} \tag{1.1}$$

Applying this to the present example, since $n_1 = 2$, $n_2 = 1$, $n_3 = 0$, $n_4 = 0$, and $N = \Sigma_i n_i = 3$ gives $W = 3!/2!1!0!0! = 3$ as already stated. (Notice that $0! = 1$.) To summarize, when three distinguishable balls are placed in four boxes with no limitations on the number per box, there are 64 possible events, whereas with the restriction that $n_1 = 2$, $n_2 = 1$, $n_3 = n_4 = 0$ there are only three possible events. Thus, if these balls are thrown at random into four boxes, *assuming that all boxes are equally accessible to all the balls,* the probability of two balls landing in box 1 and one in box 2 is 3/64.

EXAMPLE 1.2. (a) In how many ways can 12 distinguishable objects be placed in 3 boxes with 7 in the first box, 4 in the second, and 1 in the third?

(b) If the objects are thrown at random into the boxes, and if the boxes are equally accessible to the balls, what is the probability of giving the distribution in (a)?

ANSWER. (a) $W = 12!/7!4!1! = 3960$ ways; (b) $W_{tot} = 3^{12}$ so the probability is $3960/3^{12} = 0.0075$.

Instead of placing balls in boxes we now imagine that we are assigning particles (molecules or atoms) to energy levels, using the same principles. The molecules, like the balls, will be assumed to be distinguishable (for reasons to be discussed later) and the energy levels will be denoted by $\epsilon_0, \epsilon_1, \epsilon_2, \ldots, \epsilon_i$. Notice that the first level is designated 0, not 1. This is because we will usually, but not always, find it convenient to associate this level with zero energy, and a particle in this level is said to be in the ground state. Particles with energy ϵ_1, the next lowest possible energy, are said to be in the first excited state, those with energy ϵ_2 in the second excited state, and so on. In addition to the limitation that there be only N particles for distribution (because there is just so much matter available for distribution) there is a limitation on the number of particles in each level, denoted by $n_0, n_1, n_2, \ldots, n_i$ because there is a limit to the total amount of available energy, E. These limitations are described by $N = \Sigma_i\, n_i$ and $E = \Sigma_i\, n_i\epsilon_i$.

Consider, for example, the possible ways of distributing 7 distinguishable particles among 4 energy levels with energies of 0, 1, 2, and 3 equal quanta. (A quantum is simply a unit of energy, the size of which varies with the situation under discusssion.) Suppose, further, that the total energy divided

Table 1.1. Distributions for $N = 7$ and $E = 3$

Distribution	n_0	n_1	n_2	n_3	W	W/W_{tot}
(1)	6	0	0	1	7	0.08
(2)	5	1	1	0	42	0.50
(3)	4	3	0	0	35	0.42
Total					84	1.00

among the particles is only 3 quanta. Thus $N = 7$ and $E = 3$. One distribution meeting these requirements is $n_0 = 6$, $n_1 = 0$, $n_2 = 0$, and $n_3 = 1$. Another distribution which meets these requirements is $n_0 = 5$, $n_1 = 1$, $n_2 = 1$, and $n_3 = 0$; and still another is $n_0 = 4$, $n_1 = 3$, $n_2 = 0$, and $n_3 = 0$. There are no others. The number of ways in which these distributions can be realized is 7, 42, and 35, respectively, computed using (1.1), and $W_{tot} = 7 + 42 + 35 = 84$. Assuming equal accessibility of all the energy levels to all the particles, the probability of the second distribution is $42/84 = 0.50$. Table 1.1 summarizes this. The second distribution is therefore the most probable, with a probability of 0.50.

It is appropriate here to refer to another expression from combinatorial mathematics which is useful for finding W_{tot} in situations such as the one just discussed. When E *equal* quanta are distributed among N distinguishable particles

$$W_{tot} = \frac{(N + E - 1)!}{(N - 1)!E!} \tag{1.2}$$

Notice that whereas (1.1) gives the number of ways of realizing a given distribution, (1.2) gives the *total* number of ways of realizing a given system when the quanta distributed are all the same size. Applying (1.2) to the preceding example gives $W_{tot} = (7 + 3 - 1)!/(7 - 1)!3! = 84$, as already determined by a more laborious method.

Table 1.2. Distributions for $N = 50$ and $E = 5$

Distribution	n_0	n_1	n_2	n_3	n_4	n_5	W	W/W_{tot}
(1)	49	0	0	0	0	1	50	0
(2)	48	1	0	0	1	0	2,450	0
(3)	48	0	1	1	0	0	2,450	0
(4)	47	2	0	1	0	0	58,800	0.02
(5)	47	1	2	0	0	0	58,800	0.02
(6)	46	3	1	0	0	0	921,200	0.29
(7)	45	5	0	0	0	0	2,118,760	0.67
Total							3,162,510	1.00

EXAMPLE 1.3. Describe all possible distributions of 50 molecules with a total energy of five equal quanta, where the available energy levels have 0, 1, 2, 3, 4, and 5 quanta. Which distribution has the greatest W?

ANSWER. (It is helpful, in finding possible distributions, to determine first those in which the highest levels are filled as fully as possible.) There will be one distribution with $n_5 = 1$, and all the remaining molecules in the ground state; there will be another distribution with $n_4 = 1$, $n_1 = 1$, and all the other molecules in the ground state; there will be at least one with $n_3 = 1$, and so on. We will thus find the distributions shown in Table 1.2. The last distribution has the greatest W and so, if all levels are equally accessible, is the most probable distribution. Had we used (1.2) we could have found $W_{tot} = 54!/49!5! = 3{,}162{,}510$ immediately.

The reader is invited to repeat Example 1.3 for $N = 1000$ instead of 50, keeping E constant at 5 quanta. It will be found that W_{tot} is now 8.42×10^{12} and that W/W_{tot} for the most probable distribution is 0.980. We may now collect the previous results for $N = 50$ and 1000 and include the values for $N = 1{,}000{,}000$—see Table 1.3. The figures in the last column are, of course, the probabilities of the most probable distribution, within the restriction of $E = 5$ quanta. An important fact emerges: the greater the number of particles the more the probability of the most probable state approaches a dead certainty ($W/W_{tot} = 1$). In real systems we are dealing with colossal values of N (e.g., 10^{23}). We are also dealing with colossal numbers of energy levels. Suffice to say that in real systems, W for the most probable distribution—the one realizable in the greatest number of ways—is so near to W_{tot} that the probability is virtually unity. The important consequence of this is that in order to find W_{tot} for purposes which will appear later it is necessary only to find W for the most probable distribution. To this we now turn.

1.2. THE BOLTZMANN DISTRIBUTION

Given an isolated system of N identical but distinguishable particles with total energy E we now determine how this energy is distributed among the particles in the equilibrium or most probable state. The symbolism is as before: n_0 particles in the ground state with energy ϵ_0, n_1 particles with energy ϵ_1, n_2 with ϵ_2, and so on. Again, equal accessibility of energy levels to all particles is assumed. The number of particles is taken to be very large, and so is the number of energy levels, so that the magnitudes of the W's, the number of ways of realizing a given distribution, although still given by (1.1), cannot be computed as we have done above. We seek a method which finds the most probable distribution without having to find any of the less probable ones.

A new terminology will be introduced at this time. Different distributions will be referred to as *macrostates* or *config-*

Table 1.3. Dependence of W/W_{tot} on N for $E = 5$

N	W_{tot}	W/W_{tot} for largest W
50	3.16×10^{6}	0.67
1,000	8.42×10^{12}	0.98
1,000,000	8.33×10^{27}	0.99998

urations; different ways of realizing the same distribution or macrostate will be referred to as *microstates* or *complexions.* W now becomes the number of microstates for a given macrostate or the number of complexions for a given configuration. The most probable macrostate is the one with the greatest number of microstates. Macrostates are distinguishable experimentally; microstates are indistinguishable. All microstates are equally probable—it is only the overwhelmingly large number of microstates associated with the most probable macrostate which makes the latter so much more probable than any other macrostate.

It must be realized that the most probable macrostate is not a static situation. The n_0 particles with energy ϵ_0 are not the same particles from one instant to the next, nor are the n_1 particles with energy ϵ_1, etc. There must be a continual exchange of energy taking place among the particles covering all the microstates of all the macrostates over a period of time. Furthermore the particles are supposed to be independent of each other in the sense that at any one instant the energy of any one particle is independent of the energy of all the others. They are thus independent but able to exchange energy with each other. The particles in such a system are said to be weakly or loosely coupled. Finally, at least for the present, it is being assumed that there is only *one* state of a molecule with energy ϵ_0, only *one* state with energy ϵ_1, and so on. In the language of statistical mechanics all such energy levels are described as *nondegenerate.* The significance of this will become apparent later.

With a total of N particles possessing a total energy E, distributed so that there are n_0 with energy ϵ_0, n_1 with energy ϵ_1, or, in general n_i with energy ϵ_i, we have

$$N = \sum_i n_i = \text{const} \qquad (1.3)$$

$$E = \sum_i n_i \epsilon_i = \text{const} \qquad (1.4)$$

and wish to find that particular macrostate or set of n_i's for which W has the maximum value. W and the n_i's are the variables and N, E, and all the ϵ_i's are constants. It follows from (1.3) that

$$dN = \sum_i dn_i = 0 \tag{1.5}$$

and from (1.4) that

$$dE = \sum_i \epsilon_i dn_i = 0 \tag{1.6}$$

Moreover,

$$W = \frac{N!}{\prod_i n_i!} \tag{1.1}$$

Rather than maximize W we maximize $\ln W$, as it is more convenient to do so. We therefore take the natural logarithm of both sides of (1.1),

$$\ln W = \ln N! - \sum_i \ln n_i!$$

and equate the differential to zero:

$$d \ln W = d\left(\ln N! - \sum_i \ln n_i!\right) = 0 \tag{1.7}$$

We must now combine (1.5), (1.6), and (1.7). To do so we digress briefly to introduce the Stirling approximation for the natural logarithm of the factorial of a very large number, namely,

$$\ln N! = N \ln N - N \qquad (N \text{ very large}) \tag{1.8}$$

EXAMPLE 1.4. What percent error is introduced by using (1.8) to evaluate (a) ln 10! (b) ln 50!?

ANSWER. (a) Accurate calculation gives 10! = 3,628,800 and ln 10! = 15.1044. Equation (1.8) gives ln 10! = 10 × ln 10 − 10 = 13.0259, or 14% error. (b) Accurate calculation gives 50! = 3.0414×10^{64} and ln 50! = 148.478. Equation (1.8) gives ln 50! = 50 × ln 50 − 50 = 145.601, or 1.94% error. (It is clear that the approximation works better for the larger number. Actually, for $N > 100$ the error in ln $N!$ is negligible.)

We now return to the problem at hand. By taking the n_i's to be very large, and recognizing that N (and therefore ln $N!$) is constant, (1.7) becomes

$$d \ln W = -d \sum_i \ln n_i! = -d \sum_i (n_i \ln n_i - n_i)$$
$$= -d \sum_i n_i \ln n_i + d \sum_i n_i = 0$$

or

$$\sum_i (n_i \, d \ln n_i + \ln n_i \, dn_i) = 0$$

since, by (1.5),

$$d \sum_i n_i = 0$$

However, $d \ln n_i = dn_i / n_i$, so

$$\sum_i (dn_i + \ln n_i \, dn_i) = 0$$

or

$$\sum_i \ln n_i \, dn_i = 0 \qquad (1.9)$$

again using (1.5). By the method of undetermined multipliers we now combine (1.9) with (1.5) and (1.6). Equation (1.5) is multiplied by a constant, α, (1.6) is multiplied by another constant, β, and the two resulting equations are added to (1.9) to give

$$\sum_i (\ln n_i + \alpha + \beta\epsilon_i)dn_i = 0 \tag{1.10}$$

It may be noted that α must be dimensionless and β must have the dimensions of reciprocal energy to validate this procedure, since $\ln n_i$ is dimensionless.

For those familiar with the method of undetermined multipliers it is probably sufficient to say that it follows from (1.10) that $\ln n_i + \alpha + \beta\epsilon_i = 0$. We will, however, try to make this conclusion more convincing by the following argument.

Suppose, for the moment, that there are only two energy levels accessible. Then

$$(\ln n_0 + \alpha + \beta\epsilon_0)dn_0 + (\ln n_1 + \alpha + \beta\epsilon_1)dn_1 = 0$$

In this instance both dn_0 and dn_1 would have to be zero because, with both N and E fixed, only one distribution would be possible, and no change in either n_0 or n_1 could be made. (For example, if N were 3 and E were 2, n_0 would *have* to be 1 and n_1 would *have* to be 2.) Only if there were *more than* 2 levels accessible could the dn_i's be other than zero. (For example, if N were 3 and E were 2, n_0, n_1, and n_2 could be either 2, 0, and 1, or 1, 2, and 0, respectively.) In this case one of the three dn_i's could be varied independently but, having chosen it, the other two would be determined. It follows that, in the series of terms in (1.10), all but two of the dn_i's can be varied independently. Let us pick the first two, dn_0 and dn_1, as the dependent ones, and let us choose values for α and β such that both $(\ln n_0 + \alpha + \beta\epsilon_0)dn_0 = 0$ and $(\ln n_1 + \alpha + \beta\epsilon_1)dn_1 = 0$.

Since dn_0 and dn_1 are not necessarily zero the quantities in parentheses will have to be zero, enabling one to find expressions for α and β in terms of n_0, ϵ_0, n_1, and ϵ_1. These expressions will cause the first two terms in (1.10) to disappear, so that all the remaining ones must add up to zero. In these, since the dn_i's are independent and therefore can have any value, it follows that every $\ln n_i + \alpha + \beta\epsilon_i$ must be zero. Since, therefore,

$$\ln n_i + \alpha + \beta\epsilon_i = 0$$

it follows that

$$n_i = e^{-\alpha}e^{-\beta\epsilon_i} \tag{1.11}$$

Finally we evaluate α and β. To find α we note that from (1.3)

$$N = \sum_i n_i = \sum_i e^{-\alpha}e^{-\beta\epsilon_i} = e^{-\alpha}\sum_i e^{-\beta\epsilon_i}$$

since $e^{-\alpha}$ is a constant. Therefore

$$e^{-\alpha} = \frac{N}{\sum_i e^{-\beta\epsilon_i}}$$

and (1.11) becomes

$$n_i = N\frac{e^{-\beta\epsilon_i}}{\sum_i e^{-\beta\epsilon_i}} \tag{1.12}$$

Alternatively, if all the ϵ_i's are measured relative to the ground state, so that $\epsilon_0 = 0$, we can let $i = 0$ in (1.11) and find that

$$e^{-\alpha} = n_0$$

so that

$$n_i = n_0 e^{-\beta\epsilon_i} \tag{1.13}$$

The evaluation of β will not be considered until other relationships have been developed. Suffice to say here that $\beta = 1/kT$, where k is the Boltzmann constant, 1.38066×10^{-23} J/K. This will be proved in Section 4-5. Equations (1.12) and (1.13) can now be written

$$n_i = N \frac{e^{-\epsilon_i/kT}}{\sum_i e^{-\epsilon_i/kT}} \tag{1.14}$$

and

$$n_i = n_0 e^{-\epsilon_i/kT} \tag{1.15}$$

These are alternative versions of the famous Boltzmann distribution law as applied to a system of distinguishable particles occupying nondegenerate energy levels. Before examining the law more closely we abbreviate the denominator of (1.14) by the symbol q and give it the name *particle partition function* or *molecular partition function.* That is,

$$q = \sum_i e^{-\epsilon_i/kT} \tag{1.16}$$

so

$$n_i = \frac{N}{q} e^{-\epsilon_i/kT} \tag{1.17}$$

The partition function is one of the most important quantities in statistical mechanics, and will be dealt with in a separate section below and in later sections of this book. It can be evaluated, of course, when the allowed energy levels for the particle are known.

We conclude this section with a closer look at the Boltzmann distribution in the form of (1.15). First it should be observed that, at a given temperature, since all the ϵ_i's are positive, as are also k and T, the population of the levels falls off exponentially as ϵ_i increases. The ground state always has the largest population. Since nondegenerate levels have been assumed, no two levels have the same population. How rapidly the population falls off depends on how rapidly the ϵ_i's rise: a steep rise (wide spacing of the levels) causes a rapid falloff in population; a gentle rise (small spacing of the levels) causes a slow falloff. This is indicated schematically in Figure 1.1(a) and 1.1(b), where the heights of the bars are a measure of the relative populations of the levels.

It is tempting to see if the tabulated results for the two examples used earlier in establishing the meaning of W in Section 1.1 are in conformity with this exponential falloff *for the most probable distributions.* It will be seen that this is only a rough approximation. In distribution No. 7 of Example 1.3, for example, the population does indeed fall off as i increases ($n_0 = 45$, $n_1 = 5$, $n_2 = 0$), but the falloff is not exponential. The cause of the apparent failure of the Boltzmann distribution law is not hard to find: the law is valid only for very large numbers of particles, not a mere 50. It should be noted, however, that, as predicted by the law, any distribution in which the population of any energy level is greater than that of any one of lower energy (called "population inversion") is not a most probable distribution. On these grounds alone, then, we could have ruled out immediately distribution Nos. 1 to 5 as contenders for the most probable.

A second point to be made is that we can use (1.15) to compute the relative populations of any two levels in the same system at a given temperature. If a and b denote the two levels, and if (1.15) is written for both levels, and one of the two resulting expressions is divided by the other, we find that

$$n_a / n_b = e^{-(\epsilon_a - \epsilon_b)/kT} \tag{1.18}$$

It is seen that the larger the difference in the level energies, $\epsilon_a - \epsilon_b$, the greater the disparity in their populations. We may note a further corollary to this: If two levels, a, and b, are *adjacent* ones, and if all the levels are *evenly spaced*, $\epsilon_a - \epsilon_b$ becomes a constant, and so, if T is constant, n_a/n_b is constant too. This means that the ratio of the populations of any two adjacent levels is a constant—for even spacing.

EXAMPLE 1.5. For a certain system to which the Boltzmann distribution law applies the first five permitted molecular energy levels are $\epsilon_0 = 0$, $\epsilon_1 = 1.106 \times 10^{-20}$, $\epsilon_2 = 2.212 \times 10^{-20}$, $\epsilon_3 = 3.318 \times 10^{-20}$ and $\epsilon_4 = 4.424 \times 10^{-20}$ J, respectively.

(a) Find $e^{-\epsilon_i/kT}$ for each of the five levels when the system is at 300 K and at 500 K.

(b) Find $\Sigma_i e^{-\epsilon_i/kT}$ for both temperatures.

(c) What per cent of the molecules are in each level at both temperatures?

(d) What has been the effect on the population distribution of the change in temperature from 300 to 500 K?

(e) Compute the total energy of 1 mole of this system at 300 K.

(f) Find n_1/n_0, n_2/n_1, and n_3/n_2 at both temperatures.

ANSWER. (a) At 300 K, $e^{-\epsilon_0/kT} = e^0 = 1.0000$; $e^{-\epsilon_1/kT} = \exp(-1.106 \times 10^{-20}/1.381 \times 10^{-23} \times 300) = 0.0693$. Similarly the remaining terms are 0.0048, 0.0003, and 0.0000. At 500 K the respective terms are 1.0000, 0.2015, 0.0406, 0.0082, and 0.0016.

(b) $\Sigma_i e^{-\epsilon_i/kT}$ = the sum of the answers to (a), viz., 1.0744 at 300 K and 1.2519 at 500 K. (The terms beyond the fifth are negligible.)

(c) At 300 K, by (1.14), $n_0/N = 1.0000/1.0744 = 0.9307$ or 93.07%. Similarly $n_1/N = 6.45\%$, $n_2/N = 0.45\%$, and $n_3/N =$

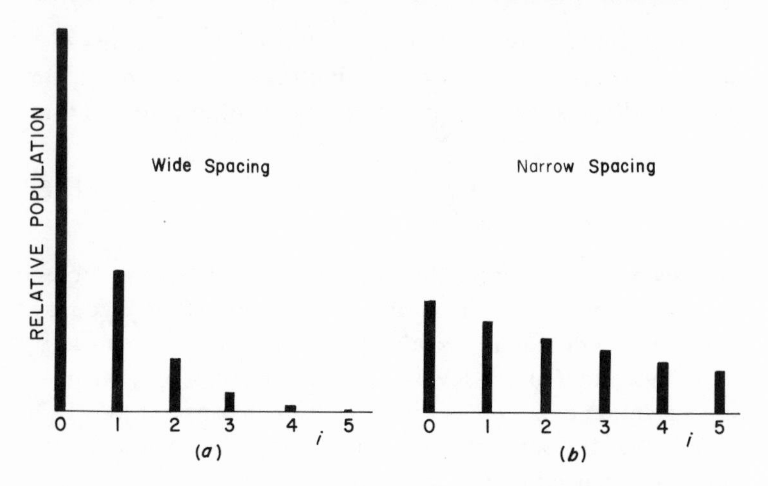

Figure 1.1. Effect of spacing on relative population of levels.

0.03%. $n_4/N \cong 0\%$. At 500 K the respective values are 79.88%, 16.10%, 3.24%, 0.66%, and 0.13%.

(d) The rise in temperature has produced a more even distribution in the population so that the falloff is more gradual. Qualitatively the difference in behavior at the two temperatures resembles the difference between Figures 1.1(a) and 1.1(b).

(e) E can be found from (1.4). Each n_i is the product of the total number of molecules (the Avogadro number, L) and the fraction in the ith level. For example, $n_0 = 0.9307L$, $n_1 = 0.0645L$, etc. Therefore $E = 0.9307L(0) + 0.0645L(1.106 \times 10^{-20}) + \cdots = 496$ J/mol.

(f) At 300 K, by (1.18), $n_1/n_0 = \exp[-(1.106 \times 10^{-20} - 0)/(1.381 \times 10^{-23} \times 300)] = 0.0693$. Similar calculations show that n_2/n_1 and n_3/n_2 also equal 0.0693. At 500 K all three ratios are equal to 0.20. The ratios are the same at any one temperature because the levels are evenly spaced.

EXAMPLE 1.6. If, as in Example 1.5, the temperature is such that the population of the first excited state is 25% of that of the ground state, what is the ratio of n_3 to n_1?

ANSWER. Since $n_1/n_0 = n_2/n_1 = n_3/n_2$ (because of the even spacing), $n_3/n_1 = (n_3/n_2)(n_2/n_1) = 0.25^2 = 0.0625$.

The Boltzmann distribution law was originally developed in the last century on the basis of classical mechanics, before quantum mechanics came into existence. Our use of "quanta of energy" and the specification that only certain energy levels be available to the particles, although an anachronism, reflects the adaptation of classical statistics to more modern concepts, and is more appropriate for our ultimate purpose.

Two important considerations must sooner or later be introduced into the calculations of W and the ensuing developments before the power of statistical mechanics in thermodynamics can be exercised to its full capacity. One is to modify the treatment to recognize the possibility of the particles being indistinguishable; the other is to recognize the possibility that two or more permitted quantum states of a particle may have the same, or very nearly the same, energy. The first consideration will be examined in a later chapter; the second is discussed immediately below.

1.3. MODIFICATIONS REQUIRED BY DEGENERACY

It is common to find that particles can exist in several different states with the same energy. A diatomic gas molecule, for example, has rotational energies given by $\epsilon_J = J(J + 1)h^2/8\pi^2 I$, where I is its moment of inertia, h is Planck's constant, and J is the rotational quantum number with only integral values. However, for every J value there are $2J + 1$ rotational quantum states with energy corresponding to that value of J. Energy levels such as these, which can be attained by *more than one* quantum state, are said to be degenerate. The general symbol g_i is used for the degeneracy of the ith level. Thus for the above rotational energy levels $g_0 = 1$, $g_1 = 3$, $g_2 = 5$, etc. Only the ground rotational level is nondegenerate in this instance. Other examples will be encountered later.

Table 1.4. Microstates for Nondegenerate Levels

ϵ_0 level	ϵ_1 level	ϵ_2 level
a,b	*c*	—
a,c	*b*	—
b,c	*a*	—

When an energy level is degenerate a particle in that level now has a choice of states, and the possible number of microstates for a given macrostate can be increased very considerably. Consider, for example, the distribution of three distinguishable particles, *a*, *b*, and *c*, among three energy levels, ϵ_0, ϵ_1, and ϵ_2, such that $n_0 = 2$, $n_1 = 1$, and $n_3 = 0$. If none of the levels is degenerate $W = 3!/(2!1!0!) = 3$, by (1.1), for this macrostate. Specifically we have the three microstates shown

Table 1.5. Microstates for Degenerate Ground Level

ϵ_0 level			
State (1)	State (2)	ϵ_1 level	ϵ_2 level
a,b	—	*c*	—
a	*b*	*c*	—
b	*a*	*c*	—
—	*a,b*	*c*	—
a,c	—	*b*	—
a	*c*	*b*	—
c	*a*	*b*	—
—	*c,a*	*b*	—
b,c	—	*a*	—
b	*c*	*a*	—
c	*b*	*a*	—
—	*b,c*	*a*	—

in Table 1.4 as shown earlier in Section 1.1. Suppose, however, that the ground level is doubly degenerate but the other two levels nondegenerate, that is, $g_0 = 2$, $g_1 = g_2 = 1$. The number of microstates for this macrostate is now much greater than 3 since the two particles in the ground level can now occupy it in four ways instead of only one. There are thus the possibilities shown in Table 1.5, or a total of 12 microstates ($W = 12$). The general expression for calculating W for distinguishable particles occupying degenerate levels is

$$W = N! \prod_i (g_i^{n_i}/n_i!) \tag{1.19}$$

Use of this expression in the above illustration would yield $W = 3!(2^2/2!)(1^1/1!)(1^0/0!) = 12$, as already found. Equation (1.19) reduces to (1.1), of course, when all the g_i's are unity.

EXAMPLE 1.7. (a) In how many ways can ten distinguishable particles be distributed among three energy levels such that $n_0 = 4$, $n_1 = 5$, and $n_2 = 1$ if the degeneracies are $g_0 = 1$, $g_1 = 2$, $g_2 = 3$?

(b) In how many ways could the distribution be made if all the levels were nondegenerate?

ANSWER. (a) $W = 10!(1^4/4!)(2^5/5!)(3^1/1!) = 120{,}960$ ways. (b) $W = 10!/(4!5!1!) = 1{,}260$ ways.

EXAMPLE 1.8. In Section 1.1 an illustration was used in which three quanta of energy were distributed among seven distinguishable particles, the energy levels being nondegenerate. Redetermine the possible distributions for $g_0 = 1$, $g_1 = 2$, $g_2 = 3$, and $g_3 = 4$, and find W for each. Which is now the most probable distribution?

ANSWER. The macrostates are the same as before insofar as the n_i's are concerned, namely, those shown in Table 1.6; but the W's are different. Distribution (3) is now the most probable.

Table 1.6. Macrostates for $N = 7$, $E = 3$ with Degeneracy

Distribution	n_0	n_1	n_2	n_3	W	W/W_{tot}
(1)	6	0	0	1	28	0.05
(2)	5	1	1	0	252	0.45
(3)	4	3	0	0	280	0.50
Total					560	1.00

The presence of degeneracy requires not only a revision of the expression for W given in (1.19) but a revision of the Boltzmann distribution law. We follow a procedure analogous to that used earlier in Section 1.2 for the nondegenerate case. With N and E still constant, and W given by (1.19),

$$dN = 0 \tag{1.5}$$

$$dE = 0 \tag{1.6}$$

$$d \ln W = d\left(\ln N! + \sum_i n_i \ln g_i - \sum_i \ln n_i!\right) = 0$$

are the three equation to be solved simultaneously. The last equation simplifies to

$$d \ln W = d\left(\sum_i n_i \ln g_i - \sum_i \ln n_i!\right) = 0$$

since N is a constant. Remembering that the g_i's are constants and applying the Stirling approximation we have, after simplification,

$$\sum_i \ln g_i \, dn_i - \sum_i \ln n_i \, dn_i - \sum_i dn_i = 0$$

or

$$\sum_i \ln (n_i/g_i)\, dn_i = 0 \tag{1.20}$$

Equation (1.20) corresponds to (1.9) in the earlier derivation.

Applying the method of undetermined multipliers leads, by the same kind of argument, to

$$n_i = N\frac{g_i e^{-\epsilon_i/kT}}{\sum_i g_i e^{-\epsilon_i/kT}} \tag{1.21}$$

and

$$n_i = \frac{n_0}{g_0} g_i e^{-\epsilon_i/kT} \tag{1.22}$$

This time the particle partition function q is given by

$$q = \sum_i g_i e^{-\epsilon_i/kT} \tag{1.23}$$

and (1.21) can be written

$$n_i = \frac{N}{q} g_i e^{-\epsilon_i/kT} \tag{1.24}$$

Equations (1.21), (1.22), and (1.24) are alternative versions of the Boltzmann distribution law, modified to allow for the possibility of degeneracy. Clearly the effect of this additional consideration has been to introduce a factor g_i before each exponential term in the expressions for n_i. Thus g_i may be thought of as a weighting factor. For this reason it is sometimes called the statistical weight of the ith level.

Finally, by writing either (1.21) or (1.22) for any two levels, a and b, and dividing one by the other we can calculate the ratio of the populations of those levels:

$$\frac{n_a}{n_b} = \frac{g_a}{g_b} e^{-(\epsilon_a - \epsilon_b)/kT} \tag{1.25}$$

analogous to (1.18).

EXAMPLE 1.9. For a certain collection of molecules at equilibrium at 100 K the energies of the first levels are 0, 2.05×10^{-22}, and 4.10×10^{-22} J. The corresponding degeneracies are 1, 3, and 5. Find the relative populations of the levels.

ANSWER. Using (1.25)

$$n_1/n_0 = (3/1)\exp[-(2.05-0)10^{-22}/1.38\times10^{-23}\times100] = 2.59$$
$$n_2/n_1 = (5/1)\exp[-(4.10-0)10^{-22}/1.381\times10^{-23}\times100] = 3.72$$

Therefore $n_1{:}n_2{:}n_3 = 1.000{:}2.587{:}3.719$. (Notice that, because of the varying degeneracy, the population does *not* necessarily fall off with rising energy. Moreover $n_1/n_0 \neq n_2/n_1$, even though the spacing of the energies is even, for the same reason.)

1.4. THE PARTICLE PARTITION FUNCTION

The quantity q, defined by (1.16) and (1.23), is one of the most important in statistical mechanics. It is well to examine it closely. Let us look first at (1.16), to which (1.23) reduces when there is no degeneracy. In less abbreviated form it is

$$q = e^{-\epsilon_0/kT} + e^{-\epsilon_1/kT} + e^{-\epsilon_2/kT} + e^{-\epsilon_3/kT} + \cdots$$

Since the ϵ_i's increase with i, and k and T are positive, the exponents of all the terms become more negative, and the exponential terms, which are always positive, become smaller and smaller, approaching zero in the limit, even though the energies become larger and larger. The entire sum is an infinite series, but a convergent one, and therefore has a finite value. It is also dimensionless. *If* the energies are all relative to the ground state, ϵ_0 is zero and the first term of the series is unity. In this event q can never be less than unity. How much larger than unity it can be depends on the spacing of the ϵ_i's and the magnitude of T.

Suppose, for purposes of illustration, that the energy spacing is even, that is $\epsilon_1 - \epsilon_0 = \epsilon_2 - \epsilon_1 = \epsilon_3 - \epsilon_2$, etc. Letting this spacing divided by kT be x, and supposing that $\epsilon_0 = 0$, we can now write $q = 1 + e^{-x} + e^{-2x} + e^{-3x} + \cdots$. If x is large, as a result of wide spacing and/or low temperature, the values of the terms in the series will fall off rapidly and q will be close to unity. Indeed, for a sufficiently large x, q will be 1.0000. On the other hand, if x is small, as a result of narrow spacing and/or high temperature, the falloff will be slow, the series will converge slowly, and q will be large. In Table 1.7 the values of the first six terms of the series are presented for $x = 10$, 1.0, and 0.01 as examples. The totals were computed

Table 1.7. Effect of Size of x on Partition Function

i	$x = 10.00$	$x = 1.000$	$x = 0.01000$
0	1.00000	1.0000	1.0000
1	0.00005	0.3679	0.9900
2	0.00000	0.1353	0.9802
3	0.00000	0.0498	0.9704
4	0.00000	0.0183	0.9608
5	0.00000	0.0067	0.9512
⋮	⋮	⋮	⋮
q = Total =	1.00005	1.5820	100.5008

using the exact relation, suitable only for this particular convergent series, $q = (1 - e^{-x})^{-1}$. The alert reader will recognize that the behavior shown here is what causes the effects shown in Figure 1.1(a) and 1.1(b). The relative magnitudes of the terms in the expressions for q are precisely what determines the relative populations of the energy levels, as seen in (1.18). Thus a small q indicates a concentration of the population in the low-energy levels, a large q a spread in population over many levels. Furthermore, if we rewrite (1.17) in the form

$$n_i/N = e^{-\epsilon_i/kT}/q \tag{1.17}$$

or (1.24) in the form

$$n_i/N = g_i e^{-\epsilon_i/kT}/q \tag{1.24}$$

we see that the fraction of the whole population which is in the ith level is given by the exponential term for that level divided by the sum of all such terms for all the levels. It is because q has these characteristics relative to how the particles are distributed or "partitioned" among the energy levels that it is called the partition function for that particle. The particles to which it usually refers are molecules; it is therefore called the molecular partition function. This must not be confused with other kinds of partition functions (for example the molar partition function) used in the statistical treatment of ensembles, and which are beyond the scope of this book.

In nondegenerate systems the terms in q diminish as i increases. For a degenerate system, however, although the series converges, the magnitude of the terms now modified by the factor g_i may increase and pass through a maximum before eventually falling off. This implies that the most populated level at equilibrium is not necessarily the one with the lowest energy.

Before leaving the general subject of partition function an alternative way of viewing or writing it, which is sometimes advantageous, should be mentioned. In writing it as we have done above, namely,

$$q = \sum_i g_i e^{-\epsilon_i/kT} \tag{1.23}$$

the summation has been over all the energy levels, any multiple quantum states with the same energy being taken care of by the degeneracy factor. It will be readily seen that we could have summed over all the quantum states instead of over all the energy levels, thus rendering the g_i's unnecessary, as follows:

$$q = \sum_{\substack{\text{quantum} \\ \text{states}}} e^{-\epsilon_i/kT} \tag{1.26}$$

We shall have occasion to use this statement to advantage in evaluating the partition function for translational energy in Chapter 5.

1.5. SUMMARY

We have in this chapter developed the elements of the statistical mechanics of systems consisting of large numbers of distinguishable particles occupying energy levels which may or may not be degenerate. The calculation of W, the number of microstates for a given macrostate, and the calculation of W_{tot}, the number of microstates for a given state, have been described, and the Boltzmann distribution law, which permits the determination of the relative population of the energy levels for the most probable macrostate, has been derived and illustrated with the help of the partition function.

This is the foundation on which to build the methods for determining the thermodynamic functions for systems of distinguishable particles, to be described in Chapter 3, with material from Chapter 2. The picture is very incomplete, however, without the statistical mechanics of indistinguishable particles. The latter is the subject matter of Chapter 4 and later chapters.

PROBLEMS

For the problems marked with an asterisk (*) only the final answers are provided at the end of the book; for those not so marked both the workings and the final answers are provided.

1.1. Derive the general expression for the number of ways of distributing N distinguishable objects in boxes such that there are n_1 in box 1, n_2 in box 2, n_3 in box 3 . . . , n_i in box i, namely, $W = N!/\Pi_i n_i!$

1.2. (a) Eight distinguishable particles comprising a non-degenerate system with a total energy of 4 quanta, occupy energy levels with 0, 1, 2, 3, and 4 quanta. Write all possible sets of n_i's and compute the number of microstates for each.

(b) Which is the most probable macrostate?

(c) If, instead of computing W for every macrostate in your search for the most probable one, you had immediately eliminated those with inverted populations, what would have been your conclusion?

1.3. As a somewhat crude approximation $N! = (N/2)^N$. Use this for all the factorial numbers in the derivation of (1.11) and determine its effect.

1.4. The famous Clausius–Clapeyron equation for the dependence of the vapor pressure, p, of a liquid with temperature, T, is $d \ln p/dT = \Delta H^v/RT^2$, where ΔH^v is the molar enthalpy of vaporization and R the gas constant. Recast it into the form of (1.15). Do the same for the isothermal dependence of barometric pressure, P, on altitude, z, namely, $dP/dz = -\rho g$, where ρ is the gas density and g the acceleration of gravity, assuming ideal gas behavior. Note what these two relations have in common with the Boltzmann distribution law.

1.5. (a) A good example of a system of distinguishable particles is a collection of independent oscillators, as will be seen in Chapter 3. Suppose that the permitted energies are given by $(v + \frac{1}{2})h\nu$, where ν is the oscillator frequency (the same for all of them) and v is an integer (0, 1, 2, 3, . . .). Letting x stand for $h\nu/kT$ the respective exponents in the expression for q become $x/2$, $3x/2$, $5x/2$, etc. Express q in closed form in terms of x with the help of the relation $\Sigma_{i=0}^{\infty} e^{-ix} = (1 - e^{-x})^{-1}$ used in Section 1-4.

(b) What is the lower limit of q?

***1.6.** The energy levels occupied by the particles of a certain substance which obeys the Boltzmann distribution law are uniformly 3.20×10^{-20} J apart, and there is no degeneracy. What fraction of the particles is in the ground level at 300 K? at 1000 K?

***1.7.** The kinetic energy of a molecule of ideal gas of mass m in a cubic container of edge a is $(n_x^2 + n_y^2 + n_z^2)h^2/8ma^2$, where n_x, n_y, and n_z are integers (quantum numbers) with possible values from 1 to infinity. Find the degeneracy of the level for which the energy is $14h^2/8ma^2$.

*1.8. (a) Calculate the particle partition function for a fictitious system of molecules at 80 K in which the accessible energies are 0, 2, 6, and 12 quanta, and the degeneracies are 1, 3, 5, and 7, respectively. Each quantum is 5.0×10^{-22} J.

(b) Which level is the most highly populated at equilibrium?

(c) What fraction of the molecules occupies the most populated level?

1.9. (a) In the following, System I has 3 quanta of energy distributed among 45 distinguishable particles and System II has 2 quanta of the same size distributed among 30 distinguishable particles. The given distributions (macrostates) are the most probable ones for both systems.

System I	$n_0 = 42$	$n_1 = 3$	$n_2 = 0$
System II	$n_0 = 28$	$n_1 = 2$	$n_2 = 0$

What is the average number of quanta per particle in each system?

(b) In neither of these distributions is the falloff in population an exponential one; that is, $n_1/n_0 \neq n_2/n_1$. Why?

(c) Imagine the systems to be combined into a single system in which 5 quanta are distributed among 75 particles. What would be the distribution *if* there were no exchange of energy between the particles of System I and System II? Find W for this situation. Assume $W_{tot} = W$ for the most probable distribution.

(d) Actually, the most probable distribution in the combined system is $n_0 = 70$, $n_1 = 5$, $n_2 = 0$. Find W for this distribution and show that an exchange of energies must have occurred when the systems were combined. [*Note:* Some aspects of this problem will be referred to in the next chapter.]

***1.10.** (a) A certain nondegenerate system consists of 25 distinguishable particles with a total energy of 5 equal quanta. The particles can have 0, 1, 2, 3, 4, or 5 quanta. How many macrostates are possible?

(b) How many microstates are possible?

(c) What is the most probable population distribution?

(d) What fraction of the total number of microstates is possessed by the most probable macrostate?

Chapter Two

THE STATISTICAL BASIS OF ENTROPY

2.1. THE BOLTZMANN–PLANCK EQUATION

Reference was made in the previous chapter to the expectation of a link between the entropy of an isolated system and its probability. On the basis that the probability of a system is measured by the number of ways it can be realized it is therefore expected that entropy, S, will be some function of W_{tot}. Although a full appreciation of this function must await consideration of systems of indistinguishable particles we will commence with those consisting of distinguishable ones.

Consider an isolated system consisting of N atoms of a pure, crystalline, monatomic solid. Although all N atoms are identical they occupy different positions in the crystal lattice and can be regarded, therefore, as distinguishable in principle. Each atom is "anchored" or "localized," but vibrating about its fixed position in the crystal lattice with energy equal to one of the many permitted energy levels. It is continually exchanging energy with the neighboring atoms but its energy at one instant is supposed to be independent of their energies. The total energy, $E = \Sigma_i n_i\epsilon_i$, is, of course, constant because the system is isolated. It is to the atoms of a system fitting the above description that the statistical mechanics developed in the preceding chapter applies.

Let us suppose we have two such systems, of exaggerated simplicity for the sake of illustration: System I, which consists of four atoms with a total energy of two equal quanta, and System II, which consists of four atoms with a total energy of three equal quanta. The atoms of System I are identical, the atoms of System II are identical, but the atoms of the two systems are not necessarily identical. The size of the quanta in the two systems are also not necessarily identical, nor is their temperature necessarily the same. Both systems are crystalline solids, so that the particles of both are regarded as distinguishable. Application of the concepts developed in Chapter 1 to these systems gives the possible macrostates for each system, and the number of microstates for each macrostate, shown in Table 2.1. The ten microstates for System I are depicted schematically in Figure 2.1(a) and the twenty microstates for System II in Figure 2.1(b), where *a, b, c,* and *d* are the System I atoms and *a′, b′, c′,* and *d′* the System II atoms, fixed in location but not in energy. In less detail the two systems may be represented as in Figures 2.2(a) and 2.2(b).

Suppose that these two systems have entropies S_I and S_{II}, respectively. We now bring them together to give a "com-

Table 2.1. Macrostates for $N = 4$, $E = 2$, and $N = 4$, $E = 3$

System I: $N = 4$, $E = 2$ quanta					
Macrostate	n_0	n_1	n_2	n_3	W
(1)	3	0	1	0	4
(2)	2	2	0	0	6
					$W_{tot} = 10 = W_I$
System II: $N = 4$, $E = 3$ quanta					
Macrostate	n_0	n_1	n_2	n_3	W
(1)	3	0	0	1	4
(2)	2	1	1	0	12
(3)	1	3	0	0	4
					$W_{tot} = 20 = W_{II}$

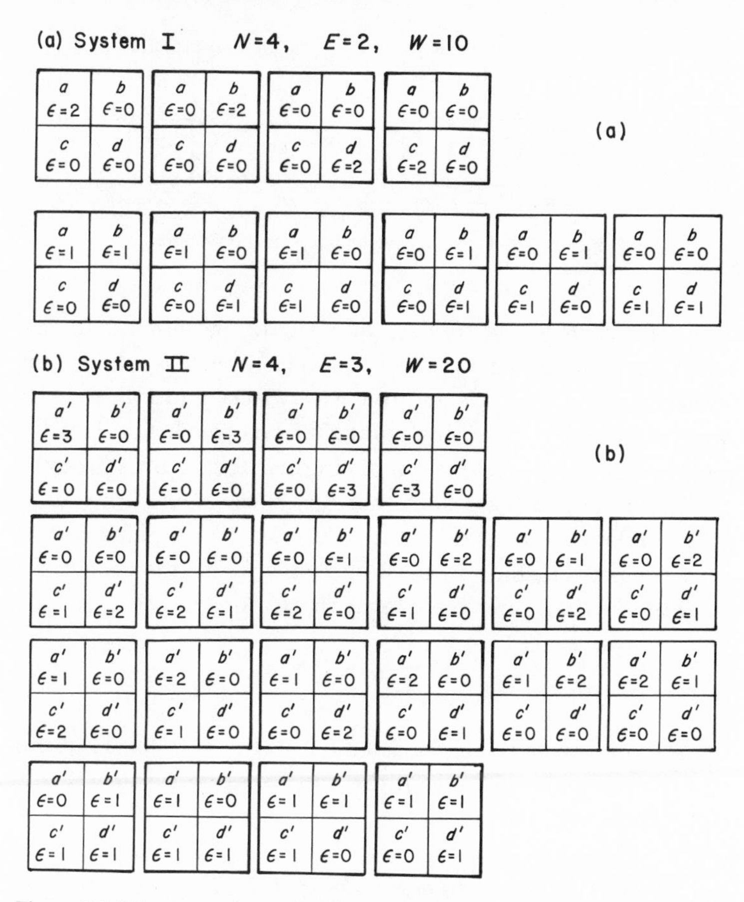

Figure 2.1. Microstates for crystalline solid with $N = 4$ and (a) $E = 2$ quanta (b) $E = 3$ quanta.

bined system" as in Figure 2.3, *keeping each isolated from the other* to prevent any possible interaction, such as the flow of energy or matter from one to the other. Now the entropy of the combined system must be $S_{\mathrm{I}} + S_{\mathrm{II}}$. However, the total number of microstates for it is $W_{\mathrm{I}} \times W_{\mathrm{II}} = 10 \times 20 = 200$,

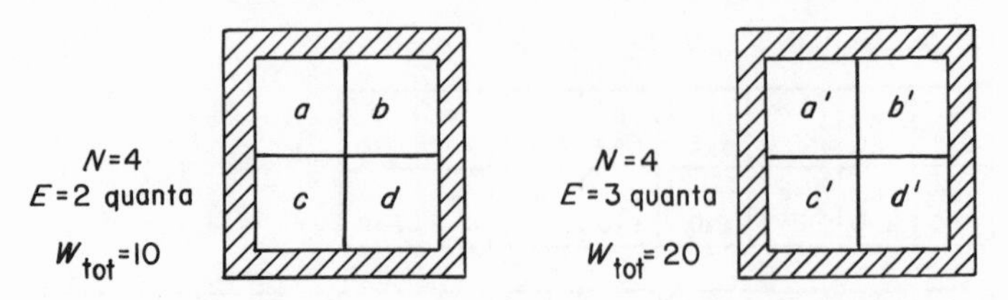

Figure 2.2. Two uncombined crystalline systems.

since for each of the 10 ways of realizing System I there are 20 ways of realizing System II. It should be clear, then, that while the entropies are additive the W's are multiplicative. It should be observed at this point that this conclusion is valid even when the particles are indistinguishable. A little thought should convince the reader that this is so. Now if the functional relations between S and W_{tot} must be the same for all three systems (I, II and the combined system) it is postulated that the function must be a logarithmic one, namely,

$$S = a \ln W_{tot} + b \tag{2.1}$$

where the constant a is the same for all three systems but the constant b could depend on the nature of the system. If this were true then

$$S_I = a \ln W_I + b$$
$$S_{II} = a \ln W_{II} + b'$$

and, adding,

$$S_I + S_{II} = a \ln (W_I \times W_{II}) + b''$$

where $b'' = b + b'$ is also a constant. Thus the functional relationship given by (2.1) would fit the requirement that S be additive and W_{tot} multiplicative.

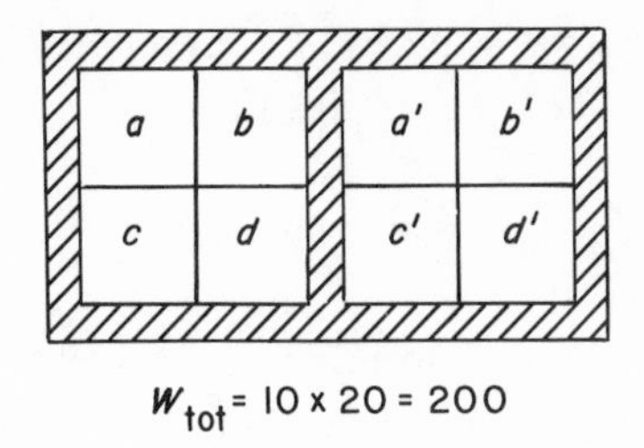

Figure 2.3. Two "combined" crystalline systems.

Equation (2.1), based on Boltzmann's concept of probability, was subsequently simplified by Planck on the basis of the following argument. According to the third law of thermodynamics a perfect crystal at the absolute zero of temperature has zero entropy. Now if systems such as those depicted in Figure 2.1 are cooled to absolute zero, *all* the atoms are expected to acquire the lowest possible energy, ϵ_0. The energy of System I is thus reduced to $E = 4\epsilon_0$ ($= 0$ if $\epsilon_0 = 0$); there is then only one macrostate ($n_0 = 4$, $n_1 = n_2 = n_3 = 0$), and there is only one microstate for that macrostate. Thus $W_{\text{tot}} = W_{\text{I}} = 1$. (We must use W_{tot}, not W for the most probable macrostate, for systems with so few particles.) Similarly System II, if cooled to absolute zero, should have only one macrostate with only one microstate, and $W_{\text{tot}} = 1$ also. Systems I and II thus become as indicated in Figure 2.4. Since $S = 0$ for both sys-

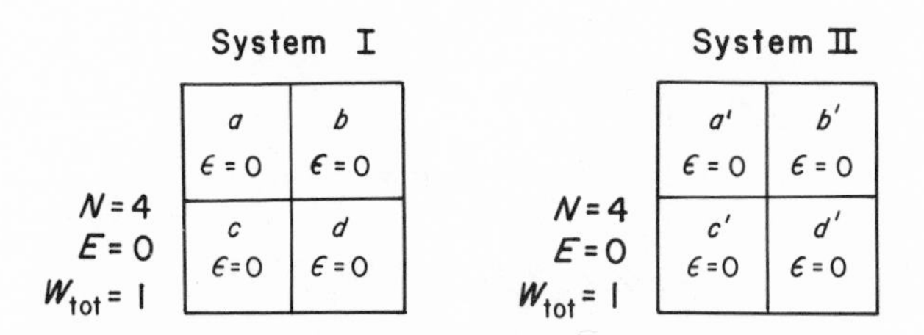

Figure 2.4. Systems of Figure 2.1 at 0 K.

tems, (2.1) becomes $0 = a \ln 1 + b$, and since $\ln 1 = 0$, b must be zero too. Thus (2.1) simplifies to

$$S = a \ln W_{tot} \tag{2.2}$$

There remains to determine the value of a. We may do this by imagining that an ideal gas is allowed to expand adiabatically into a vessel of larger volume. We let the initial volume be V_i and the final volume V_f. We visualize a thermally insulated container (Figure 2.5) of volume V_f divided into two parts by a removable partition, such that the volume on the left of it, initially containing all the gas, is V_i and that on its right, initially empty, is $V_f - V_i$. (Actually, the insulation is not really necessary because an ideal gas undergoes no energy change in free expansion.) If we had only one molecule of gas to begin with, and if V_f were twice V_i, the chance of the molecule being in volume V_i after the partition had been removed would be exactly one-half of the chance of its being in volume V_f (a dead certainty). If V_f were three times V_i the ratio of the chances would be one-third. In general, the ratio of the chances or probabilities would be V_i/V_f for this one molecule. Even if there were other molecules present the ratio of the probabilities for any given molecule would still be V_i/V_f. If there were two molecules present this would be true for each of them individually, but the ratio of the probabilities of their *both* being in the volume V_i (after the partition had been

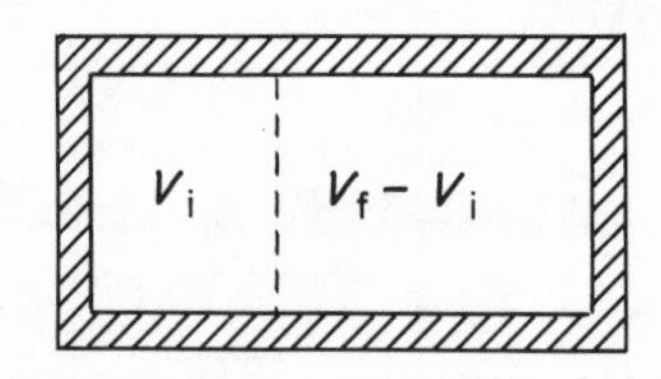

Figure 2.5. Expansion of ideal gas from *Vi* to *Vf*.

removed) to their being in the volume V_f would be $(V_i/V_f)^2$ since the probability of two events occurring simultaneously is the product of the probabilities of the individual events. It will be seen that the ratio of the probabilities is the same as the ratio of W_{tot} for the initial state to W_{tot} for the final state, which we shall write as W_i/W_f. Thus for two molecules $W_i/W_f = (V_i/V_f)^2$. For three molecules $W_i/W_f = (V_i/V_f)^3$, and for N molecules it is $(V_i/V_f)N$. Inverting this relation and replacing N by L, the Avogadro number, gives, for 1 mole of gas,

$$W_f/W_i = (V_f/V_i)^L \tag{2.3}$$

Now if (2.2) is valid, ΔS for the expansion must be $a \ln W_f - a \ln W_i$, or $a \ln (W_f/W_i)$. Combining this with (2.3) yields

$$\Delta S = a \ln (V_f/V_i)^L = aL \ln (V_f/V_i) \tag{2.4}$$

We know, however, from classical thermodynamics, that 1 mole of ideal gas suffers an increase in entropy of $R \ln (V_f/V_i)$ in expanding freely from V_i to V_f. Using this result in (2.4) gives

$$R \ln (V_f/V_i) = aL \ln (V_f/V_i) \tag{2.5}$$

It follows that $R = aL$ or

$$a = R/L \tag{2.6}$$

Thus the constant a in (2.1) and (2.2) turns out to be the gas constant per molecule. Symbolized by k and called the Boltzmann constant in his honor, it has the value 1.38066×10^{-23} J/K and is, of course, the same quantity used in Chapter 1 in connection with the denominator of the exponential terms in the partition function. It may be mentioned parenthetically that, according to Planck, Boltzmann never introduced this constant or investigated its magnitude. It is therefore possible to write (2.2) as

$$S = k \ln W_{tot} \tag{2.7}$$

This is the famous Boltzmann–Planck equation. It applies to all systems whether consisting of distinguishable or indistinguishable particles. In it S refers to the same quantity of matter as that to which W_{tot} refers. For example, if W_{tot} refers to 1 mole of substance under given conditions, S is its molar entropy, even though k is usually thought of as referring to one molecule.

It should be observed that (2.7) does not rule out the occasional appearance of less probable macrostates in an isolated system. If one reviews the arguments leading to (2.3) it will be seen that with only four molecules present, for example, and with $V_i = V_f/2$, the chance of all four molecules being in the original volume after removal of the partition is $(1/2)^4$ or $1/16$, so this event is a fairly probable one. The system will, in fact, spend one-sixteenth of the time in this configuration and not all the time in the most probable configuration. Since, however, W_{tot} remains constant in the equilibrated isolated system, regardless of brief excursions away from the most probable configuration, so does S remain constant. We hasten to add, however, that when large numbers of particles are involved—and this is always true in practice—the probability of departure from the most probable configuration is so small as to be totally negligible, as shown in Section 1.1.

EXAMPLE 2.1. Calculate the entropies of Systems I and II of Figure 2.1.

ANSWER. $S_I = k \ln 10 = 3.18 \times 10^{-23}$ J/K; $S_{II} = k \ln 20 = 4.13 \times 10^{-23}$ J/K.

The Boltzmann–Planck relation is one of the most important in all science. It expresses the concept of entropy, defined in classical thermodynamics by $dS = dq_{\text{rev}}/T$ (where q_{rev} is the heat absorbed by the system from the environment in a reversible process at temperature T), in terms of a probability. Thus it gives the statistical basis for the second law of thermodynamics. It is the link between classical and statistical thermodynamics, and reveals that, through a knowledge of the properties and energies of individual atoms and molecules, one can determine the properties of matter in the bulk. Recognition of it permits us to rewrite the second law of thermodynamics,

$$dS_{E,V} \geq 0 \tag{2.8}$$

as

$$(d \ln W_{\text{tot}})_{E,V} \geq 0 \tag{2.9}$$

which brings us back to the original premise that isolated systems approach that state with the maximum probability, W_{tot} being indistinguishable from W for the most probable configuration for large numbers of particles.

EXAMPLE 2.2. (a) Suppose that all eight atoms of the systems shown in Figure 2.3 are identical and that the barrier separating them from one another is removed so that energy (only) can pass from one to the other. The entire combined system remains isolated, however. Determine the macrostates for the resulting new system and the total number of microstates.

(b) Find the entropy of the new system and discuss the applicability of the second law of thermodynamics to the change which resulted in its formation.

ANSWER. (a) Since all the atoms are identical, the permitted energy levels are the same for both subsystems. Since, for the com-

Ludwig E. Boltzmann (1844–1906), famous physicist and founder of statistical thermodynamics, was born in Vienna and received his doctorate from the University of Vienna in 1866. He held professorships in Austria and Germany and published important theoretical papers based on the statistical approach to the interpretation of natural phenomena. He is best known for his development of the relationship governing the distribution of energy among particles of matter, and for his discovery of the connection between entropy and probability which gave a statistical interpretation to the second law of thermodynamics. The Boltzmann constant (the gas constant per molecule) was named in his honor.

Although his approach was subsequently widely accepted by the scientific world, there was much resistance and hostility toward his ideas and methods during his lifetime, and he became so depressed that he committed suicide while on a vacation in Italy. His tombstone in the Zentralfriedhof in Vienna, shown above, lies a few yards from those of Beethoven, Brahms, Mozart, Schubert, and Johann Strauss. The inscription $S = k \log W$ is mute testimony to the thermodynamic contribution of a great scientist whose life ended in sadness.

Table 2.2. Macrostates for $N = 8$ and $E = 5$

Macrostate	n_0	n_1	n_2	n_3	n_4	n_5	W
(1)	7	0	0	0	0	1	8
(2)	6	1	0	0	1	0	56
(3)	6	0	1	1	0	0	56
(4)	5	2	0	1	0	0	168
(5)	5	1	2	0	0	0	168
(6)	4	3	1	0	0	0	280
(7)	3	5	0	0	0	0	56
							$W_{tot} = 792$

bined system, $N = 8$ and $E = 5$ quanta, the macrostates and the corresponding numbers of microstates shown in Table 2.2 are possible. [Alternatively, $W_{tot} = 12!/7!5! = 792$, using (1.2).]

(b) With the barrier removed $S = k \ln 792 = 9.22 \times 10^{-23}$ J/K. Before the barrier was removed $S = k \ln 200 = 7.31 \times 10^{-23}$ J/K. Thus the entropy of the entire (isolated) system has been increased as a result of removing the barrier. This is in the direction predicted by the second law.

Note: We may suppose that, since the average energy per atom (E/N) was 0.50 and 0.75 quanta in Systems I and II, respectively, before being combined, the two systems were at different temperatures, System II being the hotter.

Equation (2.7) permits one to find the probability of a decrease in entropy in an isolated system already at equilibrium. For the latter system

$$S_{eq} = k \ln W_{eq}$$

and for some state slightly removed from the equilibrium state,

$$S' = k \ln W'$$

Subtraction of one equation from the other gives

$$S_{eq} - S' = k(\ln W_{eq} - \ln W') = k \ln (W_{eq}/W')$$

or

$$W'/W_{eq} = e^{-(S_{eq}-S')/k} \tag{2.10}$$

For example, the standard entropy of $H_2(g)$ at 298 K is 130.6 J/K mol. For some state differing from this equilibrium state by as little as 1.0×10^{-3} J/K mol to occur, (2.10) tells us that $W'/W_{eq} = \exp(-1.0\times10^{-3}/1.38\times10^{-23}) = e^{-10^{20}}$. Thus W' is such a very small fraction of W_{eq} that the chance of the nonequilibrium state occurring in an isolated system, already at equilibrium, is prohibitively small.

EXAMPLE 2.3. In Example 2.2, what is the chance of the final equilibrium system reverting to the initial configurations of Figure 2.1?

ANSWER. $W'/W_{eq} = 200/792 = 0.252$, so there is, actually, a one-in-four chance of the reappearance of the original configurations. [The same result could have been obtained using (2.10).]

Note: It is easy to see that about one-quarter of the 792 microstates in the combined system correspond to the exact sum of certain microstates in the separated systems and therefore correspond to temporary reversion to the original subsystems. For example, 16 of the 56 microstates for macrostate (3) of the combined system correspond to any one of the four microstates of macrostate (1) of System I occurring at the same instant as any one of the four microstates of macrostate (1) of System II.

2.2. FURTHER COMMENTARY ON THE BOLTZMANN–PLANCK EQUATION

The logarithmic nature of (2.7) means that W_{tot} changes much more rapidly than S. Some idea of the order of magnitude of these quantities may have been gained from Example 2.1, but let us examine a real system. The absolute entropy of NaCl(s) at 0 K is, of course, zero, whereas at 298 K it is 72.4 J/K mol. Equation (2.7) tells us that $W_{tot} = 1$ and $\exp(72.4/1.38\times10^{-23})$ or $\exp(5.25\times10^{24})$, respectively. Since W_{tot} and S are effectively measures of the same thing (since one depends only on the other) it is obviously far less cumbersome to work with S than with W_{tot}. Nevertheless it is W_{tot} which is at the heart of the statistical thermodynamic method and which, at least in principle, must be determined before determining S. How this is done will be shown in the next and subsequent chapters.

A popular view of entropy is that it measures the amount of disorder or chaos in a system. If this is so then W_{tot} is also a measure of disorder, and naturally occurring processes in isolated systems must proceed in the direction of increasing disorder. A word of caution, however, is appropriate in this usage, as will be evident from the following illustration: Consider a supercooled liquid which, while placed in isolation transforms spontaneously (at least partly) into the corresponding crystalline solid. That such a change is spontaneous will not be denied. But is the resulting solid more disordered than the liquid from which it formed? In reconciling what appears superficially to be a contradiction of the second law it must be remembered that the freezing process causes heat to be evolved and, since the system is isolated, its temperature will rise, so that the solid formed will be hotter than the original supercooled liquid. The combined solid and unfrozen liquid which result are both at a higher temperature than the original and, in this state, W_{tot} for the system has a larger value.

2.3. DEPENDENCE OF W_{tot} ON E AND ON V

The frequent reference in the foregoing sections to isolated systems is to ensure that they have a fixed number of atoms or molecules, N, occupy a fixed volume, V, and have a fixed energy, E. When N is fixed the thermodynamic state functions are determined by V and E. The entropy is in this category so $S = f(E,V)$ for a fixed N. However, since S and W_{tot} are closely linked through (2.7), it can be seen that W_{tot} is also a function of E and V for a fixed N. The nature of these dependences will now be examined.

Now we know from classical thermodynamics, that, for a *closed* system at equilibrium, and in equilibrium with the environment,

$$TdS = dE + PdV \tag{2.11}$$

as long as the only kind of work performed by the system is mechanical or p-V work. Eq. (2.11) has been called the fundamental equation of thermodynamics and represents, in differential form, a combination of the first and second laws. It follows that

$$\left(\frac{\partial S}{\partial E}\right)_V = \frac{1}{T} \tag{2.12}$$

and

$$\left(\frac{\partial S}{\partial V}\right)_E = \frac{P}{T} \tag{2.13}$$

Introducing (2.7) gives the dependence of W_{tot}, or rather of ln W_{tot} on E and on V:

$$\left(\frac{\partial \ln W_{tot}}{\partial E}\right)_V = \frac{1}{kT} \tag{2.14}$$

$$\left(\frac{\partial \ln W_{tot}}{\partial V}\right)_E = \frac{P}{kT} \tag{2.15}$$

Just as (2.12) may be thought of as a classical definition of temperature so (2.14) may be thought of as a statistical definition. Somewhat analogously (2.15) may be thought of as a statistical definition of pressure. We note from (2.12) that, since T is positive, S increases with E, and from (2.13), that S increases with V under the given restrictions. Similarly, by (2.14) and (2.15), W_{tot} increases with E and with V under the given restrictions.

The operation of (2.14) has, in fact, already been noted. If we return once again to the two systems of Figure 2.1 it will be recalled that $N = 4$, $E = 2$ in I, and $N = 4$, $E = 3$ in II. Let us assume that Systems I and II refer to the same four atoms of a crystalline solid occupying the same volume, so that II has arisen only by increasing E of I at constant volume. Since W_I and W_{II} were found to be 10 and 20, respectively, it is evident that increasing E at constant volume has increased W_{tot}, as required by (2.14).

An alternative way of looking at this is to recall that, as T increases, the spacing between adjacent energy levels divided by kT, to which the symbol x was given in Section 1.4, decreases. This causes the population to spread out over more of the higher levels and so increases the number of occupied levels. This, in turn, increases W_{tot} and ln W_{tot} because of the increased number of choices of levels for each particle.

Another implication of (2.14) is that the higher the temperature the smaller the value of $1/kT$ and the less sensitive ln W_{tot} is to changes in E. To illustrate this consider two more

Table 2.3. Effect of Temperature Change

System	N	E	W_{tot}	$\ln W_{tot}$	E/N	
I	4	2	10	2.30	0.50	Average: 0.6
II	4	3	20	3.00	0.75	
III	4	6	84	4.43	1.50	Average: 1.6
IV	4	7	120	4.79	1.75	

systems, III and IV. In System III, $N = 4$ and $E = 6$ equal quanta giving $W_{tot} = 84$, and in System IV, $N = 4$ and $E = 7$ equal quanta giving $W_{tot} = 120$. [These may readily be computed using (1.2).] If all four systems, I, II, III, and IV, refer to the same four atoms in the same volume with the same size quanta, they differ only in the number of quanta of energy. The average E per atom, E/N, may be taken as indicative of the temperature of each system. We therefore have the values shown in Table 2.3, so that the order of increasing temperature is I, II, III, IV. From I and II $(\partial \ln W_{tot}/\partial E)_V \cong (\Delta \ln W_{tot}/\Delta E)_V = (3.00 - 2.30)/1 = 0.70$ whereas from III and IV it is $(4.79 - 4.43)/1 = 0.36$. It is thus seen that, at a temperature such that $E/N \cong 0.6$, $\ln W_{tot}$ changes by 0.70 per added quantum, whereas at a higher temperature ($E/N \cong 1.6$) it changes by only about 0.36 per added quantum. The rise in temperature has diminished the sensitivity of $\ln W_{tot}$ to changes in E.

PROBLEMS

2.1. (a) When 10^{-10} mol of ideal gas occupies a 10 m^3 vessel what is the probability that it will all collect momentarily in a particular 9 m^3 of this space?

*(b) Repeat the calculation for 10^{-22} mol.

***2.2.** (a) The entropy of a certain crystalline solid is 3.68 ×

10^{-4} J/K mol at 1 K. Find W_{tot} and compare it with W_{tot} at 0 K.
(b) What is the chance that the entropy will be only 3.67 $\times$ 10^{-4} J/K mol at 1 K?

2.3. (a) The standard entropy of graphite at 298, 410, and 498 K is 5.69, 9.03, and 11.63 J/K mol, respectively. If one mole at 298 K is surrounded by thermal insulation and placed next to one mole at 498 K, also insulated, how many microstates are there altogether for the combined but independent systems?
(b) If the same two samples are now placed in thermal contact and brought to thermal equilibrium the final temperature will be 410 K. (The final temperature is not exactly the average of 298 and 498 K because the heat capacity increases with temperature.) How many microstates are there now in the combined system? Neglect any volume changes.
(c) Apply (2.9) as a criterion of spontaneity.

2-4. When 1 mole of supercooled liquid water at 263 K in a sealed, insulated vessel eventually reaches its most stable state it is found that 0.875 mol $H_2O(l)$ and 0.125 mol ice, both at 273 K, are present. Ignoring the (very small) effects of changing pressure within the vessel, use the following data to show that the entropy of the system has undergone an increase and therefore is in accordance with the second law of thermodynamics. Enthalpy of fusion of ice = 6025 J/mol at its melting point (273 K); heat capacity of $H_2O(l)$ at constant pressure = 75.3 J/K mol (independent of temperature over a small range).

***2-5.** How does the fractional or relative change in W_{tot} depend on E at constant volume for a given quantity of substance?

2-6. (a) A system of N localized oscillators with E equal quanta has one additional quantum added to it at constant volume. With the help of the Boltzmann–Planck equation show that the resulting increase in entropy is given approximately by $k \ln [(N + E)/E]$. Any reasonable approximations may be made.

(b) Use (2.12) to express T in terms of N, E, and the energy of one quantum.

Chapter Three

THERMODYNAMIC FUNCTIONS FOR SYSTEMS OF LOCALIZED (DISTINGUISHABLE) PARTICLES

3.1. THE DISTRIBUTION LAW

In the preceding chapter we have laid the groundwork for determining all the state functions of a system of distinguishable particles. We proceed immediately to erect a structure on this groundwork by showing that all these functions can be expressed in terms of the partition function. Then, as soon as the latter can be calculated, all the state functions can also be found.

We begin with the energy, E, and combine (1.4) and (1.24) to give

$$E = \sum_i n_i \epsilon_i = \frac{N}{q} \sum_i \epsilon_i g_i e^{-\epsilon_i/kT} \tag{3.1}$$

bearing in mind that N and q are constant for a system of fixed N and fixed V. In order to replace the Σ with something more convenient we note that, since

$$q = \sum_i g_i e^{-\epsilon_i/kT} \tag{1.23}$$

$$\left(\frac{\partial q}{\partial T}\right)_V = \sum_i (\epsilon_i/kT^2) g_i e^{-\epsilon_i/kT} = (1/kT^2) \sum_i \epsilon_i g_i e^{-\epsilon_i/kT}$$

so (3.1) can be written

$$E = \frac{N}{q} kT^2 \left(\frac{\partial q}{\partial T}\right)_V$$

which is the same as

$$E = NkT^2 \left(\frac{\partial \ln q}{\partial T}\right)_V \tag{3.2}$$

This is more convenient for purposes of calculation than (3.1). Furthermore, since the heat capacity at constant volume, C_V, is defined by

$$C_V = (\partial E/\partial T)_V \tag{3.3}$$

it follows that

$$C_V = \left[\frac{\partial}{\partial T} NkT^2 \left(\frac{\partial \ln q}{\partial T}\right)_V\right]_V = Nk \left[\frac{\partial}{\partial T} T^2 \left(\frac{\partial \ln q}{\partial T}\right)_V\right]_V \tag{3.4}$$

Moreover, since the enthalpy of the system, H, is defined by

$$H = E + PV \tag{3.5}$$

we can write

$$H = NkT^2 \left(\frac{\partial \ln q}{\partial T}\right)_V + PV \tag{3.6}$$

and also write the heat capacity at constant pressure, C_P, defined by

$$C_P = (\partial H/\partial T)_P \tag{3.7}$$

as

$$C_P = \left\{ \frac{\partial}{\partial T} \left[NkT^2 \left(\frac{\partial \ln q}{\partial T} \right)_V + PV \right] \right\}_P \tag{3.8}$$

The remaining state functions, entropy, Helmholtz free energy, and Gibbs free energy are obtained through the Boltzmann–Planck equation,

$$S = k \ln W_{\text{tot}} \tag{2.7}$$

Recalling that, for a macroscopic system, $W_{\text{tot}} = W$ for the most probable macrostate, and that, for the latter

$$W = N! \prod_i (g_i^{n_i}/n_i!) \tag{1.19}$$

where the n_i's are for the most probable state, we see that

$$\begin{aligned} S &= k \ln \left[N! \prod_i (g_i^{n_i}/n_i!) \right] \\ &= k \ln N! + k \sum_i n_i \ln g_i - k \sum_i \ln n_i! \end{aligned} \tag{3.9}$$

Application of the Stirling approximation to this yields

$$\begin{aligned} S &= kN \ln N - kN + k \sum_i n_i \ln g_i - k\left(\sum_i n_i \ln n_i - \sum_i n_i \right) \\ &= kN \ln N - k \sum_i n_i \ln (n_i/g_i) \end{aligned} \tag{3.10}$$

199167

since $\sum_i n_i = N$. Now, since the n_i's are those for the most probable distribution, they must be those given by the Boltzmann distribution law, (1.24), which can be written

$$\ln (n_i/g_i) = \ln (N/q) - \epsilon_i/kT \tag{3.11}$$

on taking natural logarithms. Insertion of this into (3.10), and recalling (1.3) and (1.4), yields

$$\begin{aligned} S &= kN \ln N - k \sum_i n_i \left(\ln \frac{N}{q} - \frac{\epsilon_i}{kT} \right) \\ &= kN \ln N - k \sum_i n_i \ln \frac{N}{q} + k \sum_i \frac{n_i \epsilon_i}{kT} \\ &= kN \ln N - kN \ln \frac{N}{q} + \frac{E}{T} \\ &= kN \ln N - kN \ln N + kN \ln q + \frac{E}{T} \end{aligned}$$

or

$$S = kN \ln q + \frac{E}{T} \tag{3.12}$$

For one mole of substance (3.12) becomes

$$S = R \ln q + \frac{E}{T} \tag{3.13}$$

Since, in (3.2), E has already been expressed in terms of q, we now have S also in terms of q.

We can use (3.12) to express the Helmholtz free energy, A, defined by

$$A = E - TS \tag{3.14}$$

as follows:

$$A = E - T\left(kN \ln q + \frac{E}{T}\right)$$

or

$$A = -kNT \ln q \tag{3.15}$$

Finally, the Gibbs free energy, G, defined by

$$G = A + PV$$

becomes

$$G = -kNT \ln q + PV \tag{3.16}$$

Thus we have succeeded in expressing E, C_V, H, C_P, S, A, and G as functions of the partition function. There remains to examine how it may be computed for a system of localized particles.

3.2. ATOMIC CRYSTALS—THE EINSTEIN MODEL

A system of distinguishable, localized particles, the subject of the present chapter, is exemplified by a crystalline, "atomic" solid such as silver or diamond. As outlined in Section 2.1 the atoms vibrating about their individual fixed positions are assumed to be loosely coupled, that is, to be independent of each other in the sense that the energy of each one at any instant is independent of that of its neighbors, even though there are forces of attraction and a continual exchange of energy among them. This would not only be assumed to

LORETTE WILMOT LIBRARY
NAZARETH COLLEGE

apply to metallic elements, the atoms of which are held together by metallic bonds, but to molecular crystals such as diamond, where strong covalent forces are present.

It is further assumed that the overall vibration is effectively a three-dimensional one with no one direction favored over any other, that is, isotropic. We may think of this as the result of the superposition of three one-dimensional oscillations, each of which is simple harmonic, meaning that the restoring force is proportional to the displacement from the position of rest. Thus a sample consisting of N atoms has $3N$ oscillations associated with it.

Einstein assumed that the frequency, ν, of these oscillations is *the same* for all the atoms in the sample. This assumption, combined with quantum mechanical concepts, suggests that the permitted energies for a one-dimensional oscillator be used for all of the $3N$ oscillations, namely,

$$\epsilon_v = (v + \tfrac{1}{2})h\nu \qquad (3.17)$$

where v is a vibrational quantum number with integral values 0, 1, 2, . . . , etc., and h is Planck's constant (6.626×10^{-34} J s). When $v = 0$, $\epsilon_v = \epsilon_0 = \tfrac{1}{2}h\nu$, the zero-point energy. (Since $\tfrac{1}{2}h\nu \neq 0$ the atom still has some vibrational energy, even though it is in its lowest vibrational state. An atom with no vibrational energy would be one completely at rest in the equilibrium position in the crystal lattice—a hypothetical condition—otherwise described as being at the bottom of the potential energy well.) Thus (3.17) gives the vibrational energy levels with energy zero equal to that in the hypothetical motionless condition, and $\epsilon_0 = \tfrac{1}{2}h\nu$, $\epsilon_1 = \tfrac{3}{2}h\nu$, $\epsilon_2 = \tfrac{5}{2}h\nu$, etc. The atoms, all with the same frequency, differ only with respect to the number of quanta they possess, the latter being distributed according to the Boltzmann distribution law. At some other temperature the distribution of quanta would, of course, be different, but the frequency would be the same at all temperatures. It should be recalled, however, that q, given by (1.24), and as used in the previous section, is taken to depend not only on T but, in principle, on the volume V (and therefore on the pressure). There is no provision, however, in the simple Einstein model for a volume dependence of q, it being assumed, because of the small compressibility of solids, that the dependence is small enough to ignore. Some of these assumptions are open to challenge, but are reasonable, so we shall see whether they lead to predictions that are in accord with experiment.

It should be added that the only form of energy which the atoms are assumed to possess is vibrational energy, and that there is no degeneracy in the ground electronic or nuclear states. The significance of this will be seen in a later chapter.

Following the procedure suggested in Section 3.1 we first determine the partition function, q_v, where the subscript indicates that the kind of energy with which we are concerned is vibrational. Combining (3.17) with the definition of q in (1.24), and taking from quantum mechanics the fact that the energies of one-dimensional oscillators are nondegenerate, gives

$$q_v = e^{-h\nu/2kT} + e^{-3h\nu/2kT} + e^{-5h\nu/2kT} + \cdots \tag{3.18}$$

This is now abbreviated by writing θ for $h\nu/k$, θ being called the (Einstein) characteristic temperature, since it does have the dimensions of a temperature and depends on ν which is characteristic of the given substance:

$$q_v = e^{-\theta/2T} + e^{-3\theta/2T} + e^{-5\theta/2T} + \cdots$$

which can be written

$$q_v = e^{-\theta/2T}(1 + e^{-\theta/T} + e^{-2\theta/T} + e^{-3\theta/T} + \cdots)$$

(The reader is reminded that the energy zero has been taken to be the hypothetical atom at rest in its equilibrium position in the crystal lattice. We need not have made this choice. Had we chosen some other energy zero, such as the ground vibrational state, many of the subsequent developments would not have been altered. More will be said about this later.) The expression in parentheses is, conveniently, equal to $(1 - e^{-\theta/T})^{-1}$ (see Section 1.4), as will be confirmed in Problem 3.1, so the expression for q_v becomes

$$q_v = e^{-\theta/2T}/(1 - e^{-\theta/T}) \tag{3.19}$$

We are now in a position to calculate the state functions for an Einstein solid provided ν, which leads to θ, is known. Since ν is a property of a particle it means that we are on our way to determining properties of matter in the bulk from the properties of its individual particles.

The first state function we shall calculate is the energy, E, of one mole of solid. We do this by substituting (3.19) into (3.2) after replacing N in the latter by $3L$ (because each atom is equivalent to three oscillators). This gives

$$\begin{aligned}
E &= 3LkT^2 \left\{\frac{\partial}{\partial T} \ln\,[e^{-\theta/2T}/(1 - e^{-\theta/T})]\right\}_V \\
&= 3RT^2 \left\{\left[\frac{\partial}{\partial T}(-\theta/2T)\right]_V - \left[\frac{\partial}{\partial T} \ln\,(1 - e^{-\theta/T})\right]_V\right\} \\
&= 3RT^2\,[(\theta/2T^2) + (\theta/T^2)e^{-\theta/T}/(1 - e^{-\theta/T})] \\
&= \frac{3}{2}R\theta + \frac{3R\theta\, e^{-\theta/T}}{1 - e^{-\theta/T}}
\end{aligned}$$

since θ is independent of T.

Finally, dividing both numerator and denominator of the second term by $e^{-\theta/T}$ gives

$$E = \frac{3}{2} R\theta + \frac{3R\theta}{e^{\theta/T} - 1} \tag{3.20}$$

which is the vibrational energy of one mole of an Einstein solid with energy zero equal to the hypothetical motionless oscillator.

EXAMPLE 3.1. Taking ν for lead to be $1.9 \times 10^{12}\ \mathrm{s}^{-1}$, find the vibrational energy per mole of lead at 300 K (with energy zero at the bottom of the potential well). Assume an Einstein solid.

ANSWER. The value of θ for lead must be $6.63 \times 10^{-34} \times 1.9 \times 10^{12}/1.38 \times 10^{-23} = 91$ K, and at 300 K $\theta/T = 0.30$. Therefore $E = (3/2)(8.31 \times 91) + 3(8.31 \times 91)/(e^{0.30} - 1) = 7620$ J/mol.

It is clear from (3.20) that as T approaches zero E approaches $(3/2)R\theta$, since $e^{\theta/T}$ approaches infinity, and, of course, all the atoms are then in the ground state and possess only the zero-point energy. If E_0 denotes this vibrational energy, that is, writing

$$E_0 = \tfrac{3}{2} R\theta \tag{3.21}$$

(3.20) becomes

$$E - E_0 = \frac{3R\theta}{e^{\theta/T} - 1} \tag{3.22}$$

On the other hand, as T becomes indefinitely large, E approaches

$$\lim_{T\to\infty} \left[\frac{3}{2} R\theta + \frac{3R\theta}{e^{\theta/T} - 1}\right]$$

$$= \lim_{T\to\infty} \left[\frac{3}{2} R\theta + \frac{3R\theta}{1 + \theta/T + (1/2!)(\theta/T)^2 + \cdots - 1}\right]$$

the expanded denominator reduces to θ/T, and the whole expression reduces to $\frac{3}{2}R\theta + 3R\theta/(\theta/T) = E_0 + 3RT$, so $E - E_0$ approaches $3RT$.

With respect to (3.20) it is probably obvious that if our choice of energy zero had been the ground vibrational state (see Problem 3.3) the expression for the vibrational energy would have been given by

$$E = \frac{3R\theta}{e^{\theta/T} - 1} \tag{3.23}$$

instead of by (3.20), since E_0 in (3.21) and (3.22) is zero. Proof of this from first principles will be included in the problems at the end of the chapter.

For (3.20) and (3.23) it is apparent that, in general, the vibrational energy of an Einstein solid is not directly proportional to E at constant volume, except at high temperatures where $E - E_0 = 3RT$, provided $E_0 = 0$. This may be contrasted with the energy of an ideal monatomic gas where the energy, all kinetic, *is* proportional to T.

EXAMPLE 3.2. (a) Consider 1 mole of each of two Einstein solids, A and B. The oscillator frequency in A is $15 \times 10^{12}\ \mathrm{s}^{-1}$ and that in B is twice that in A. If the energy zero is taken as the zero point energy in both solids, and if B has the same total number of quanta of energy as has A, how does the average energy of an A atom compare with that of a B atom?

(b) How do the temperatures of A and B compare?

ANSWER. (a) If n stands for the number of quanta in A and in B, $E_A = nh\nu_A$ and $E_B = nh\nu_B$. Since $\nu_B = 2\nu_A$, $E_B = 2E_A$, and since there are the same number of atoms in A and B the average energy of a B atom must be twice that of an A atom.

(b) Since $\nu_B = 2\nu_A$, $\theta_B = 2\theta_A$. For E_B to be $2E_A$ while $\theta_B = 2\theta_A$ it is apparent from (3.23) that θ/T must remain constant. Therefore $T_B = 2T_A$.

Note: In this example E is directly proportional to T because E was changed by changing the size of the quanta, *keeping the total number of quanta fixed.* This must not be confused with the effect of changing E without changing the size of the quanta. It the latter case, E is *not* directly proportional to T.

With an expression for the vibrational energy now determined for an Einstein solid we may find one for C_V using (3.20) or (3.22) in (3.3):

$$C_V = \left(\frac{\partial E}{\partial T}\right)_V = 3R\theta\left[\frac{\partial}{\partial T}(e^{\theta/T} - 1)^{-1}\right]_V = -3R\theta\,\frac{e^{\theta/T}(-\theta/T^2)}{(e^{\theta/T} - 1)^2}$$

or

$$C_V = 3R(\theta/T)^2\,\frac{e^{\theta/T}}{(e^{\theta/T} - 1)^2} \tag{3.24}$$

It is at this point that we can begin to see whether the results are borne out by experiment. We note, first, that according to (3.24) C_V approaches zero as T approaches zero. This may be shown by expanding each exponential using $e^x = 1 + x^2/2! + x^3/3! + \cdots$ to give

$$C_V = 3R(\theta/T)^2\,\frac{1 + (1/1!)(\theta/T) + (1/2!)(\theta/T)^2 + (1/3!)(\theta/T)^3 + \cdots}{[1 + (1/1!)(\theta/T) + (1/2!)(\theta/T)^2 + (1/3!)(\theta/T)^3 + \cdots - 1]^2} \tag{3.25}$$

Now as T approaches 0 all the terms in θ/T in the expanded numerator are so much greater than unity that the numerator becomes the same as the quantity which is squared in the denominator, and the expression reduces to

$$\begin{aligned} C_V &= 3R(\theta/T^2)/\,[(1/1!)(\theta/T) + (1/2!)(\theta/T)^2 \\ &\quad + (1/3!)(\theta/T)^3 + \cdots] \\ &= 3R/\,[(1/1!)(\theta/T)^{-1} + (1/2!) + (1/3!)(\theta/T) \\ &\quad + (1/4!)(\theta/T)^2 + \cdots] \end{aligned}$$

which approaches zero as T approaches 0, since θ/T approaches infinity. This conclusion is amply confirmed by experiment for all crystalline solids, whether monatomic or not.

At the other extreme (3.25) predicts that C_V should approach $3R$ as T approaches infinity: With T very large the expanded numerator in (3.25) reduces to unity and the expanded denominator to $[(1/1!)(\theta/T)]^2$, so that C_V approaches $3R(\theta/T)^2[1/(\theta/T)]^2$ or $3R$. This is also confirmed experimentally. Even at room temperature C_V is close to $3R$ (24.942 J/K mol) for many metallic elements, and is the basis for the nineteenth century Dulong and Petit rule, according to which the product of atomic weight and specific heat (which is the same thing as the heat capacity per mole) is approximately 6 cal/K mol or 25 J/K mol for solid elements. The Dulong and Petit rule is, therefore, actually a limiting law, valid at sufficiently high temperatures. In many instances room temperature is high enough for this limiting value to be reached. It may also be observed that a value of $3R$ for C_V is just what was predicted by the classical equipartition theory which required a contribution of $k/2$ for every "square term" needed to describe the energy of a particle. Since an atom of a solid has three such terms to describe its potential energy, and another three to describe its kinetic energy, C_V per atom should be $6(k/2) = 3k$, or $3R$ per mole.

Inspection of (3.24) shows that for a monatomic solid to attain a given value of C_V the required T will have to be greater the larger its θ value, that is, the larger the oscillation frequency of its atoms. Since high-atomic-weight elements such as lead have generally lower frequencies than low-atomic-weight elements like diamond, the latter must be at a higher temperature than the former if they are to have the same C_V. Thus, in general, one finds that C_V rises more slowly with T for those elements of lower atomic weight.

It should be pointed out that specific heats or heat capacities for solids are invariably measured at constant pressure, whereas (3.24) is for constant volume. While C_P and C_V are not greatly different they often differ at room temperature by as much as 5%. As 0 K is approached the difference vanishes. $C_P - C_V$ can be found from the exact relation

$$C_P - C_V = \alpha^2 VT/\beta \tag{3.26}$$

where α, the coefficient of thermal expansion, is $(1/V)(\partial V/\partial T)_P$, and β, the isothermal compressibility, is $-(1/V)(\partial V/\partial T)_T$. [Equation (3.26) comes from classical thermodynamics.] Thus, in principle, all C_P's should be changed to C_V's before comparison with (3.24).

EXAMPLE 3.3. Given that, for Zn(s), $\alpha = 8.93 \times 10^{-5}\ \text{K}^{-1}$, $\beta = 1.5 \times 10^{-6}\ \text{atm}^{-1}$, atomic weight = 65.4 g/mol, density = $7.14 \times 10^3\ \text{kg/m}^3$, and C_P = 24.6 J/K mol at 25°C and 1 atm, find C_V at the same temperature. (1 m^3 atm = 1.01×10^5 J.)

ANSWER. Let C_P, C_V, and V all refer to 1 mole. Then $V = 65.4 \times 10^{-3}/7.14 \times 10^3 = 9.16 \times 10^{-6}\ \text{m}^3/\text{mol}$. Therefore, using (3.26), $C_P - C_V = 24.6 - C_V = (8.93 \times 10^{-5})^2(9.16 \times 10^{-6})(298)(1.01 \times 10^5)/1.5 \times 10^{-6} = 1.47$ J/K mol, so $C_V = 23.1$ J/K mol.

Note: The energy conversion factor must not be forgotten. It would have been unnecessary had β been expressed in SI units, namely, m^2/Newton.

Alternatively one may use the Nernst–Lindemann relation:

$$C_P - C_V = AC_P^2 T \tag{3.27}$$

where the constant A, given by

$$A = \alpha^2 V/\beta C_P^2 \tag{3.28}$$

has the advantage of being nearly independent of T. Thus (3.27), although less rigorously true than (3.26), permits finding C_V from C_P knowing α and β at a different temperature.

EXAMPLE 3.4. (a) For Ag(s) at 293 K, atomic weight = 108 g/mol, $\alpha = 5.83 \times 10^{-5}\ \mathrm{K}^{-1}$, $\beta = 1.0 \times 10^{-6}\ \mathrm{atm}^{-1}$, density = 10.5×10^3 kg/m^3, and C_P = 25.46 J/K mol. Find the value of the constant A in (3.27).

(b) Find C_V for Ag(s) at 100 K if C_P at that temperature is 20.17 J/K mol.

ANSWER. (a) The molar volume is $108 \times 10^{-3}/10.5 \times 10^3 = 1.03 \times 10^{-5}$ m^3/mol. Therefore, using (3.28), $A = (5.83 \times 10^{-5})^2(1.03 \times 10^{-5})\,(1.01 \times 10^5)/(1.0 \times 10^{-6})(25.46)^2 = 5.4 \times 10^{-6}$ mol/J.

(b) Using (3.27), $20.17 - C_V = 5.4 \times 10^{-6}(20.17)^2(100) = 0.22$ J/K mol or C_V = 19.95 J/K mol.

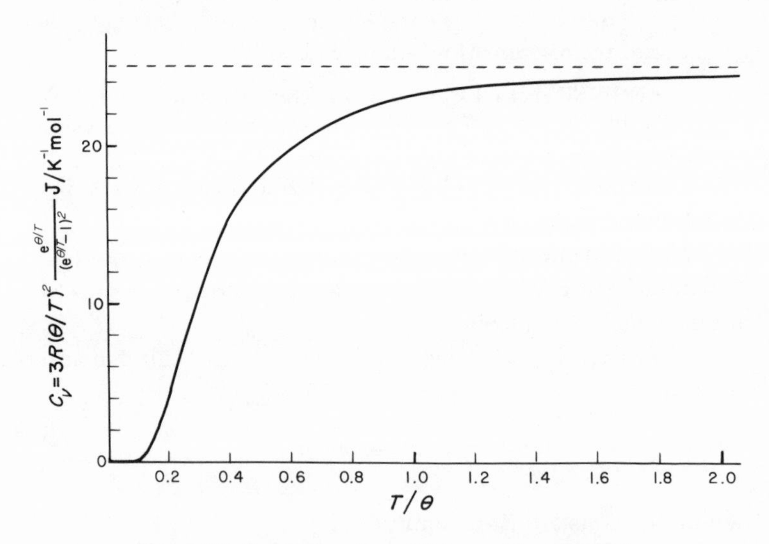

Figure 3.1. C_V as a function of T for an Einstein solid.

TABLE 3.1. C_V as a Function of θ/T for an Einstein Solid

θ/T	∞	10.00	5.00	2.00	1.25	0.50	0
T/θ	0	0.10	0.20	0.50	0.80	2.00	∞
C_V (J/K mol)	0	0.113	4.26	18.06	21.93	24.43	24.94

Returning to (3.24) we see that it predicts that C_V should be zero at 0 K and increase to a value of $3R$ at sufficiently high temperature. If valid, it predicts that C_V should be the *same* function of θ/T for all Einstein solids. This function is plotted in Figure 3.1. Typical values of C_V for various values of θ/T and T/θ are given in Table 3.1. In order to predict values of C_V, however, a value of θ is needed. This is done by taking a single measurement of C_P at some temperature, preferably where it is rising rapidly, converting it to C_V, and determining from Figure 3.1 the corresponding value of T/θ, from which θ is then found. With it, values of C_V at any other temperature can then be calculated using (3.24), and compared with experiment.

As an example let us choose silver, for which C_V [found from C_P using (3.26) or (3.27)] is 20.09 J/K mol at 103.14 K. Reference to a plot such as Figure 3.1 tells us that T/θ must be $0.614 = 103.14/\theta$, so θ must be 168 K. Using this value in (3.24) for various T's gives the calculated values of C_V shown in Table 3.2. Included are the experimental C_V's for comparison.

TABLE 3.2. Comparison of Calculated and Experimental Values of C_V

T (K)	20	30	40	50	100	200	300
C_V(calc) (J/K mol)	0.40	2.91	6.80	10.50	19.82	23.53	24.30
C_V(expt) (J/K mol)	1.67	4.81	8.49	11.80	19.87	23.39	24.19

It is seen that the experimental results are reproduced quite well at the higher temperatures, but that the predicted ones are too low at the lower temperatures. This is true in general for all monatomic solids, and indicates some flaw in the Einstein treatment, even though the overall reproduction of the experimental results for C_V is impressively good. In seeking to find where any error might lie in the assumptions, one may wonder whether it was correct to assume that the particles are really independent, whether the vibrations are really isotropic, whether the atoms do vibrate with a single frequency, whether this frequency is independent of volume, etc. We shall postpone further comment on this, but it is worth pointing out that one of the triumphs of the Einstein model was its prediction that C_V should be a universal function of θ/T for all monatomic solids. This means, of course, that any two Einstein solids, under conditions where they have the same value of θ/T, must have the same C_V. Under these conditions the two are said to be in corresponding states. For example, Ag(s), $\theta = 168$ K, at 100 K and diamond, $\theta = 1364$ K at 812 K are said to be in corresponding states because θ/T ($= 168/100 = 1364/812 = 1.68$) is the same for both, and C_V is the same for both (within the approximations of the Einstein treatment).

3.3. THE DEBYE MODEL

The inability of the Einstein model to reproduce C_V quantitatively, particularly at lower temperatures, led Debye a short time later to try to improve it. Realizing that the vibrating atoms are not really independent but are strongly coupled, he treated the crystal as an elastic medium with a *range* of frequencies, ν, varying from zero to a maximum value, ν_D, characteristic of the crystal and defined such that the total number of modes equals $3N$. In a manner analogous to Einstein's approach we let $\theta_D = h\nu_D/k$, called the Debye temperature. Similarly, $x = h\nu/kT = \theta/T$ and $x_D = h\nu_D/kT = \theta_D/T$. The details of Debye's treatment are beyond the scope of this book, but some of the results are of interest and will be presented briefly. An important integral, needed in the evaluation of thermodynamic properties, is D, which is a function of θ_D/T and is known as a Debye function. This dimensionless integral is defined as follows:

$$D = 3\left(\frac{T}{\theta_D}\right)^3 \int_0^{x_D} \frac{x^3}{e^x - 1}\,dx \tag{3.29}$$

It cannot be reduced to a simpler expression and must be evaluated numerically. Representative values are given in Table 3.3. Some of the thermodynamic functions can be expressed in terms of D as follows:

$$E - E_0 \text{ (per mole)} = 3RTD \tag{3.30}$$

$$C_V \text{ (per mole)} = 3R\left[4D - \frac{3x_D}{e^{x_D} - 1}\right] \tag{3.31}$$

and

$$S \text{ (per mole)} = 4RD - 3R\ln\left(1 - e^{-x_D}\right) \tag{3.32}$$

TABLE 3.3. Debye Heat Capacity Function[a]

θ_D/T	D	θ_D/T	D	θ_D/T	D	θ_D/T	D
0.0	1.0000	4.0	0.1817	8.0	0.03656	12.0	0.01125
1.0	0.6744	5.0	0.1176	9.0	0.02620	13.0	0.00886
2.0	0.4411	6.0	0.07758	10.0	0.01930	14.0	0.00710
3.0	0.2836	7.0	0.05251	11.0	0.01457	15.0	0.00577

[a]Defined by (3.29).

These are only the contributions from vibration and do not include, for example, any electronic contributions. It is to be noted that (3.31) shows that C_V is a universal function of θ_D/T, analogous to Einstein's conclusion that it is a universal function of θ/T, so that a plot of C_V vs. T/θ_D should be valid for all solids fitting the Debye theory. Such a plot is similar to Figure 3.1. From a single measurement of C_V at a known temperature the corresponding value of T/θ_D can be read, and therefore θ_D found. This permits the calculation of $E - E_0$, C_V, and S by means of (3.30), (3.31), and (3.32) after D has been determined from Table 3.3. (More extended tables are available in reference works, the use of which reduces the interpolation errors.)

EXAMPLE 3.5. Given that $\theta_D = 315$ K for Cu(s) find C_V at 100 K.

ANSWER. At 100 K, $\theta/T = 315/100 = 3.15$. Linear interpolation of Table 3.3 gives $D = 0.268$. Using this value in (3.31) leads to $C_V = 3R\,[(4)(0.268) - 3(3.15)/(e^{3.15} - 1)] = 16.2$ J/K mol. (More detailed tables of D show that its value is 0.2653, giving $C_V = 15.92$ J/K mol.)

The limiting values of C_V at both ends of the temperature scale are of interest. As T increases indefinitely (3.31) shows that C_V approaches $3R$, as found with the Einstein theory and by experiment. As T approaches zero the familiar "T-cubed law" emerges:

$$\lim_{T \to 0} C_V \text{ (per mole)} = \frac{12\pi^4}{5} R \left(\frac{T}{\theta_D}\right)^3 \tag{3.33}$$

where $C_V \propto T_3$. This is widely used to extrapolate heat capacity data beyond the lowest temperature of measurement to 0 K. Note that at such low temperatures the difference between C_P

and C_V is negligible, as seen from (3.26). The law appears to apply to many crystalline compounds, not just to atomic crystals.

Heat capacities calculated by (3.31) are more in accord with experiment at lower temperatues than are those calculated by the Einstein result, (3.24), and are generally more satisfactory, although not entirely so. This may be illustrated by computing C_V for Ag(s) at various temperatures using (3.31). With $\theta_D = 215$ K the values given in Table 3.4 are obtained. D was found from tables which are more detailed than Table 3.3. The last row of data is taken from Table 3.2. It will be seen that the agreement with experiment, particularly at the lower temperatures (where the Einstein theory was less successful), is good, but at the higher temperatures the Debye values are a little too large. However, the overall improvement of the Debye theory over that of Einstein is evident.

A representative calculation of entropy by means of the Debye theory is given in Example 3.6. Again the approximate nature of the results is evident.

EXAMPLE 3.6. If θ_D is 390 K for Al(s) find the molar entropy at 200 K according to the Debye theory.

ANSWER. Since $x_D = \theta_D/T = 1.95$, D from Table 3.3 is 0.453. Equation (3.32) gives $S = 4R(0.453) - 3R \ln(1 - e^{-1.95}) = 18.9$ J/K mol. [The third law value is 19.13 J/K mol. The agreement is better than would have been found by means of the Einstein theory (see Problem 3.10) but the result must still be regarded as only an approximation.]

TABLE 3.4. Heat Capacity of Silver (Debye)

T (K)	20	30	40	50	100	200	300
D	0.01560	0.04928	0.10016	0.1592	0.4133	0.6539	0.7670
C_V (calc) (J/K mol)	1.54	4.50	8.13	11.46	20.02	23.56	25.34
C_V (expt) (J/K mol)	1.67	4.81	8.49	11.80	19.87	23.39	24.19

Among the reasons which may be cited for the shortcomings of the Debye approach are the choice of permitted frequencies, the neglect of the expansion of the solid with rise in temperature, and, for metals, the neglect of the contribution of the free electrons.

PROBLEMS

3.1. Show by "long division" that, as used in Sections 1.4 and 3.2, $(1 - e^{-x})^{-1} = 1 + e^{-x} + e^{-2x} + e^{-3x} + \cdots$.

3.2. In Section 3.2 we have treated N three-dimensional oscillators as the equivalent of $3N$ nondegenerate one-dimensional oscillators, each with permitted energies given by $\epsilon_i = (i + \frac{1}{2})h\nu$. An alternate approach is to treat them as three-dimensional oscillators from the outset, with energies $\epsilon_i = (i + \frac{3}{2})h\nu$, but each having a degeneracy $g_i = \frac{1}{2}(i + 1)(i + 2)$. Show that this leads to the same result as (3.20) for E per mole.

3.3. Write out the vibrational partition function for an atom of an Einstein solid with energy zero equal to the $v = 0$ state, and use it to derive (3.23) for one mole.

3.4. Draw a schematic graph of E vs T for an Einstein solid with the help of (3.3) and Figure 3.1.

3.5. (a) For Cu(s) $C_V = 16.11$ J/K mol at 100 K. Assuming it to be an Einstein solid find the characteristic temperature. Use the latter to estimate C_V at 298 K.
(b) Convert the C_V just found to C_P at the same temperature using (3.26) given that $\alpha = 4.95 \times 10^{-5}\ \text{K}^{-1}$, $\beta = 7.50 \times 10^{-7}\ \text{atm}^{-1}$, density $= 8.93 \times 10^3\ \text{kg/m}^3$ (all at room temperature), and atomic weight $= 63.5$ g/mol. ($1\ \text{m}^3\ \text{atm} = 1.013 \times 10^5$ J.)

(c) Compare the calculated C_P with the experimental value of 24.5 J/K mol.

***3.6.** For Pb(s) at 298 K, $\alpha = 8.82 \times 10^{-5}\ \mathrm{K}^{-1}$, $\beta = 2.42 \times 10^{-6}\ \mathrm{atm}^{-1}$, atomic weight = 207.2 g/mol, density = $11.35 \times 10^3\ \mathrm{kg/m^3}$, and C_P = 26.84 J/K mol. If C_V at 150 K, calculated by theory, is 24.50 J/K mol estimate the theoretical value of C_P at that temperature. ($1\ \mathrm{m^3\ atm} = 1.013 \times 10^5$ J.)

3.7. (a) If the Einstein characteristic temperatures for boron and beryllium are 1000 and 770 K, respectively, what can be said about the relative strengths of the interatomic forces in these two crystalline solids?

(b) For which element is C_V expected to rise the more rapidly with increase in T?

***3.8.** Consider 1 mole of each of two Einstein solids, A and B. The oscillator frequency in B is 50% greater than that in A, and C_V for A at 100 K is 7.60 J/K mol. At what temperature will C_V for B have this value?

***3.9.** Combine (3.19) and (3.20) with (3.12) to find an expression for the molar entropy of an Einstein crystal in terms of θ and T, remembering that there are L atoms but $3L$ oscillations.

***3.10.** Use the result of Problem 3.9 to estimate the molar entropy of Al(s) at 200 K according to the Einstein theory, given that θ = 260 K. (The third law entropy is 19.13 J/K mol.)

3.11. (a) The Einstein theory makes no provision for the dependence of ν and therefore of θ, E, C_V, and S on volume or pressure at a given temperature. If, at 298 K, the pressure on a mole of diamond is increased from 1 to 16,000 atm, find the actual change in the entropy from the following data: atomic weight = 12.0 g/mol, density = 3.51×10^3 kg/m^3, $\alpha = 7.3 \times 10^{-5}$ K^{-1}. (Hint: Recall that $(\partial S/\partial P)_T = -(\partial V/\partial T)_P$ from classical thermodynamics.)

(b) If S^0_{298} for diamond is 2.439 J/K mol, by what per cent is its entropy changed?

3.12. When equal amounts of thermal energy are added, at constant volume, to a mole of A and a mole of B, initially at the same temperature, the latter will rise more in the substance with the larger oscillator frequency; that is, C_V will be less for that substance with the higher-frequency atomic vibrations. Both A and B are Einstein solids. Rationalize this result using (2.14).

3.13. (a) Imagine two Einstein solids, A and B, each with 100 atoms, both solids being at 0 K. What is W_{tot} for both systems? What is S for both systems?

(b) The size of the quanta required to excite the A atoms is 6.6×10^{-22} J, to excite the B atoms is 13.2×10^{-22} J. If 13.2×10^{-22} J of energy is added at constant volume to A and an equal amount to B, describe the new microstates and find the new W_{tot} for each system.

(c) In which system has W_{tot} increased by the greater amount?

(d) Which system has undergone the greater increase in entropy?

(e) In which system has the rise in temperature been greater? Justify your answer.

3.14. Given that the Einstein characteristic temperature for beryllium is 770 K calculate its molar entropy at 616 K by two methods: (1) using (3.12) and (2) using (2.7) after first computing W_{tot}. Treat the system as containing $3L$ nondegenerate one-dimensional oscillators. In connection with the second method use W_{tot} as given by (1.1) and find the n_i's by means of the Boltzmann distribution law, (1.17). Consider only the first eight energy levels and use the Stirling approximation wherever possible.

***3.15.** For Cu(s), $\theta_D = 315$ K. Predict the molar values for $E - E_0$, C_V, and S at 315 K.

3.16. (a) For diamond, $\theta_D = 1860$ K. What value does the Debye theory predict for its molar entropy at 298.15 K? (The literature value is 2.439 J/K mol.)

(b) What is unusual about the result? Account for this unusual feature.

Chapter Four

SYSTEMS OF NONLOCALIZED (INDISTINGUISHABLE) PARTICLES

4.1. THE DISTRIBUTION LAW

In Chapter 1 we developed the elementary statistical mechanics of systems of distinguishable particles, and in Chapter 3 we showed how the concepts could be used to determine the thermodynamic functions for such systems, exemplified by the Einstein and Debye models for crystalline solids. In this chapter we consider the same aspects of systems of indistinguishable particles. This is in order to be able eventually to tackle the subject of gases, the molecules of which, because of their translational motion, must be regarded as indistinguishable.

In Section 1.1 we discussed the number of ways in which seven distinguishable particles, with a total energy of three quanta, could be distributed among four energy levels with energies of 0, 1, 2, and 3 equal quanta, respectively. The results were as shown in Table 4.1. Suppose, however, that the particles are *in*distinguishable. The 7 microstates for macrostate (1) are no longer distinguishable, nor are the 42 microstates of macrostate (2) or the 35 of (3). The macrostates are still distinguishable, however. Thus the W_{tot} reduces from 84 to 3 when we remove the property of distinguishability. At first sight this may appear to be a simplification of the problem, but it so

TABLE 4.1. Macrostates for $N = 7$ and $E = 3$

Macrostate	n_0	n_1	n_2	n_3	W
(1)	6	0	0	1	7
(2)	5	1	1	0	42
(3)	4	3	0	0	35
					$W_{tot} = 84$

happens that when we apply the concept of indistinguishability to gas molecules the existence of a very considerable degeneracy introduces complications.

It will be recalled that the presence of degeneracy required modifying the original Boltzmann distribution law to give (1.24), for example, and in Problem 1.7 a brief reference was made to the degeneracy of the kinetic energy levels of an ideal gas. Let us examine the latter further.

According to quantum mechanics a molecule with mass m possessing only kinetic energy, and allowed to move in only one dimension, is permitted to have energies given by $\epsilon = n^2h^2/8ma^2$, where a is the length of the one-dimensional container and $n = 1, 2, 3, \ldots$. (A value of zero for n is forbidden because the molecule is then not even in the container!) This is the famous "particle in a box" problem. For a two-dimensional box two quantum numbers are needed, and, for a three-dimensional one, three are needed. Since the shape of the container has no bearing on the properties of the gas molecules one may choose a cubic container with edge a. The permitted energies are now given by

$$\epsilon_i = (n_x^2 + n_v^2 + n_z^2)h^2/8ma^2 \tag{4.1}$$

where n_x, n_y, and n_z can have integral values, and all three cannot be zero. For, say, the average H_2 molecule at 298 K the kinetic energy is $\frac{3}{2}kT = 6.17 \times 10^{-21}$ J. We can use (4.1) to obtain some idea of the magnitude of the factor in parentheses. Putting $\epsilon_i = 6.17 \times 10^{-21}$ J, $h = 6.63 \times 10^{-34}$ J s, $m = 2.0 \times 10^{-3}/L = 3.32 \times 10^{-27}$ kg and, say, $a = 10^{-1}$ m (that is, the gas is in a 1-liter vessel), gives $n_x^2 + n_y^2 + n_z^2 = 3.73 \times 10^{18}$. The degeneracy of this level is the number of combinations of three integers the sum of the squares of which equals this very large number. Since a very large number of such combinations is possible the energy level has a very large degeneracy. For any other substance the value will be even larger. We see, then, that for translational energy levels, that is, for kinetic energies of motion of molecules from one place to another between collisions, the degeneracy of a particular level is a very large number under ordinary conditions—in fact very much larger than the number of particles in that level.

It should be pointed out here that there are two kinds of statistics of indistinguishable particles: Fermi–Dirac and Bose–Einstein. In Fermi–Dirac statistics a given quantum state may be vacant or have only single occupancy; in Bose–Einstein a given quantum state can be occupied by any number of particles. Fermions, particles following Fermi–Dirac statistics, contain an odd number of elementary particles with half integral spins whereas bosons, those following Bose–Einstein statistics, contain an even number.

The difference between these two types may be illustrated by answering the simple question: In how many ways can two identical (indistinguishable) molecules *(a, a)* be placed in a given energy level which has a multiplicity of four; that is, when $n_i = 2$ and $g_i = 4$ for a particular i value. For bosons there are the following ten ways of doing this: *aa*, —, —, —; —, *aa*, —, —; —, —, *aa*, —; —, —, —, *aa*; *a*, *a*, —, —; *a*, —, *a*, —; *a*, —, —, *a*; —, *a*, *a*, —; —, *a*, —, *a*; —, —, *a*, *a*. (Note that the interchange of particles does not produce a new microstate because of their indistinguishability.) For fermions, on the other hand, there are only six ways of meeting the requirements, namely, the last six of the ten given for bosons.

In general, W for bosons can be calculated from the expression $(g_i + n_i - 1)!/(g_i - 1)!n_i!$ and W for fermions from $g_i!/n_i!(g_i - n_i)!$. Thus, in the above example, $W = (4 + 2 - 1)!/(4 - 1)!2! = 10$ and $W = 4!/2!(4 - 2)! = 6$, respectively. These expressions pertain to a particular energy level. Since there is a multitude of such levels we may write

$$W_{\text{bosons}} = \prod_i \frac{(g_i + n_i - 1)!}{(g_i - 1)!n_i!} \tag{4.2}$$

and

$$W_{\text{fermions}} = \prod_i \frac{g_i!}{n_i!(g_i - n_i)!} \tag{4.3}$$

The reader is reminded that the n_i's in (4.2) and (4.3) are normally those for the most probable distributions, so that the W calculated thereby is also that for the most probable distribution. However, as shown in Chapter 1, this W is virtually indistinguishable from the W for all possible distributions (W_{tot}).

Now as pointed out above, g_i in practice is very much larger than n_i. Imposing this restriction ($g_i \gg n_i$) in both (4.2) and (4.3) reduces them *both* to

$$W_{\text{tot}} = \prod_i \frac{g_i^{ni}}{n_i!} \tag{4.4}$$

so it turns out, fortunately, that it is unnecessary to determine whether Fermi–Dirac or Bose–Einstein statistics is followed for ideal gas molecules. It will also be evident that, since $g_i \gg n_i$, it is highly unlikely that more than one particle will occupy a given quantum state, so that single occupancy or no occupancy is virtually guaranteed, even if not required. Equation (4.4) for indistinguishable particles is the counterpart of (1.19) for distinguishable ones. Clearly they differ by the factor $N!$

EXAMPLE 4.1. (a) A fictitious system obeying Fermi–Dirac statistics has three indistinguishable particles with a total of three quanta of energy. The permitted energy levels have 0, 1, 2, and 3 quanta with the following degeneracies: $g_0 = 1$, $g_1 = 3$, $g_2 = 4$, $g_3 = 6$. Describe all the macrostates which are possible.

(b) Find the total number of microstates for each macrostate using (4.3).

(c) Why does (4.4) not give the correct answers to (b)?

ANSWER. (a) See Table 4.2. Macrostate (1) is ruled out because there is only one quantum state in the ground level, so only one of the two particles can be accommodated. The 12 microstates for macrostate (2) result from placing the one particles in each of the three quantum states of the one-quantum level for every time the one particle in the two-quantum energy level is placed in each of its four quantum states. In macrostate (3) the three particles in the one-quantum level are in different quantum states.

(b) For macrostate (2) $W = (1!/0!1!)(3!/3!0!)(4!/0!4!)(6!/0!6!) = 12$; for macrostate (3) $W = (1!/0!1!)(3!/3!0!)(4!/0!4!)(6!/0!6!) = 1$.

(c) Use of (4.4) would have given $W = 12$ for macrostate (1)—which is correct by accident; $W = 4\frac{1}{2}$ for macrostate (2)—obviously incorrect. Equation (4.4) is invalid because the g_i's are *not* much larger than the n_i's in this simple example.

TABLE 4.2. Macrostates for Fermi–Dirac System

Macrostate	n_0 $g_0 = 1$	n_1 $g_1 = 3$	n_2 $g_2 = 4$	n_3 $g_3 = 6$	W
(1)	2	0	0	1	Ruled out
(2)	1	1	1	0	12
(3)	0	3	0	0	1
					Total = 13

We are now ready to derive the distribution law for systems of indistinguishable particles for which $g_i \gg n_i$, using (4.4). We adopt the same procedure as in Chapter 1 by finding what set of n_i's gives the maximum W in an isolated system, that is, with both N and E constant. This involves setting $d \ln W$ equal to zero. By comparison of (4.4) with (1.19) we see that, on taking the logarithm of both expressions for W and then differentiating, *both* of them reduce to

$$d \ln W = d\left(\sum_i n_i \ln g_i - \sum_i \ln n_i!\right) = 0$$

There is, consequently, no difference between what was done in deriving (1.22) and deriving what is needed here—the results are the same. We have, therefore, even for systems of indistinguishable particles,

$$n_i = \frac{n_0}{g_0} g_i e^{-\epsilon_i/kT} \tag{4.5}$$

and

$$n_i = \frac{N}{q} g_i e^{-\epsilon_i/kT} \tag{4.6}$$

where, as before,

$$q = \sum_i g_i e^{-\epsilon_i/kT} \tag{4.7}$$

Equation (1.25) applies here, too, of course.

It is perhaps surprising that the same distribution law holds whether the particles are distinguishable or indistinguishable, provided $g_i \gg n_i$. It means that the classical Boltzmann statistics applies to both kinds of systems (if $g_i \gg n_i$) even when it is necessary to take the results of quantum mechanics for the permitted energy levels.

4.2. CALCULATION OF THERMODYNAMIC FUNCTIONS

In the preceding chapter it was shown how, for systems of distinguishable particles, the Boltzmann distribution law and Boltzmann–Planck equation could be combined with the expression for W_{tot}, Equation (1.1), to express all the thermodynamic functions in terms of the partition function. For systems of indistinguishable particles the procedure is the same, but the expression for W_{tot} is different, which leads to some important differences.

Since it is still true, according to (3.1), that

$$E = \frac{N}{q} \sum_i \epsilon_i g_i e^{-\epsilon_i/kT} \tag{4.8}$$

it follows that

$$E = NkT^2 \left(\frac{\partial \ln q}{\partial T} \right)_V \tag{4.9}$$

which is the same as (3.2). Moreover, since

$$C_V = (\partial E/\partial T)_V \tag{4.10}$$

we can write

$$C_V = Nk \left[\frac{\partial}{\partial T} T^2 \left(\frac{\partial \ln q}{\partial T} \right)_V \right]_V \tag{4.11}$$

and

$$H = NkT^2 \left(\frac{\partial \ln q}{\partial T} \right)_V + PV \tag{4.12}$$

and

$$C_P = (\partial H / \partial T)_P \tag{4.13}$$

and

$$C_P = \left\{ \frac{\partial}{\partial T} \left[NkT^2 \left(\frac{\partial \ln q}{\partial T} \right)_V + PV \right] \right\}_P \tag{4.14}$$

all of which are the same as found for distinguishable particles.

With functions that involve entropy, however, the results are not the same. Combination of (4.4) with the Boltzmann–Planck equation gives

$$S = k \left(\sum_i n_i \ln g_i - \sum_i \ln n_i! \right) \tag{4.15}$$

which differs from the analogous (3.9) in lacking the term in $N!$. Application of the Stirling approximation gives

$$S = k \sum_i (n_i \ln g_i - n_i \ln n_i + n_i) = k \sum_i \left(n_i \ln \frac{g_i}{n_i} + n_i \right)$$

or

$$S = -k \sum_i n_i \ln \frac{n_i}{g_i} + kN \tag{4.16}$$

Now (1.24), the Boltzmann distribution law, can be written as

$$\ln \frac{n_i}{g_i} = \ln \frac{N}{q} - \frac{\epsilon_i}{kT}$$

and substitution of this into (4.16) yields

$$S = -k \sum_i n_i \left(\ln \frac{N}{q} - \frac{\epsilon_i}{kT} \right) + kN$$

$$= -k \left(\sum_i n_i \right) \ln \frac{N}{q} + \frac{1}{T} \sum_i n_i \epsilon_i + kN$$

which is the same as

$$S = kN \ln \frac{q}{N} + \frac{E}{T} + kN \tag{4.17}$$

For one mole of particles this becomes

$$S = R \ln \frac{q}{L} + \frac{E}{T} + R \tag{4.18}$$

When the right sides of (3.13) and (4.18) are compared it is seen that the latter is less than the former by $R(\ln L - 1)$, a positive quantity. Thus, other things being equal, the introduction of the property of indistinguishability has reduced the molar entropy by that amount. This is in the expected direction since W_{tot} is also less (by a factor of $L!$).

EXAMPLE 4.2. (a) For a certain system of indistinguishable molecules at 300 K and 1 atm, $q = 1.00 \times 10^{30}$ and its calculated energy above ground is 3740 J/mol. Find its entropy per mole.

(b) What would have been its calculated entropy per mole had the molecules been distinguishable?

ANSWER. (a) By (4.18) $S = R \ln (1.00 \times 10^{30}/6.02 \times 10^{23}) + 3740/300 + R = 139.9$ J/K mol.

(b) If the molecules had been distinguishable the calculated entropy, however, would have been given by (3.13), or $S = R \ln (1.00 \times 10^{30}) + 3740/300 = 586.8$ J/K mol [which is, of course, $R(\ln L - 1)$ or 446.9 J/K mol greater than the result in (a)].

Since the Helmholtz and Gibbs free energies are entropy dependent we can expect the property of indistinguishability to affect them too. Accordingly, $A = E - TS$ or

$$A = E - \mathrm{T}\left[kN \ln \frac{q}{N} + \frac{E}{T} + kN\right]$$

so

$$A = -kNT \ln \frac{q}{N} - kNT \tag{4.19}$$

and, similarly, since $G = A + PV$

$$G = -kNT \ln \frac{q}{N} - kNT + PV = -kNT \ln \frac{q}{N} \tag{4.20}$$

for ideal gases.

It is worthwhile to recall that various choices for energy zero are available, particularly when dealing with vibrational energy, and the choice will determine the numerical value of the partition function, as shown in Chapter 3, for example. *If the lowest accessible energy is assigned a value of zero,* that is, if $\epsilon_0 = 0$, all the ϵ_i's will be values "above ground" and the E's, H's, A's, and G's so calculated will also be values "above ground." To ensure that this is understood it is helpful to think of E, H, A, and G in (4.9), (4.12), (4.19), and (4.20) as really being $E - E_0$, $H - H_0$, $A - A_0$, and $G - G_0$, respectively. They are often so written.

4.3. IDEAL GAS MOLECULES AND THE FACTORIZATION OF THE PARTITION FUNCTION

For the remainder of this chapter, indeed for the remainder of this book, we shall focus our attention mostly on the statistical thermodynamics of ideal gases. This restriction is not so severe as it may seem at first because most gases are nearly ideal under ordinary conditions. Because of the molecular motion the molecules must be treated as indistinguishable and, as shown earlier, have energies which are so highly degenerate as to make the Boltzmann law valid. We restrict ourselves to ideal gases in order to eliminate potential energy contributions—the influence on the energy of a molecule of the presence of its neighbors. (It will be recalled that this restriction was also made in Chapter 3 in discussing solids, but there the validity of the restriction was more questionable.) There are, however, a number of other forms of energy: translational (or kinetic—the energy of motion in straight lines between collisions), rotational (or tumbling), and vibrational (the various kinds of oscillations of the atoms of the molecules with respect to each other) being the principal ones. In addition, molecules can possess electronic energy, and their nuclei can possess nuclear energy, but these can often be conveniently ignored, and we shall do so until forced to consider them. The extension of the following treatment to include them when needed will be obvious.

It must be emphasized that all these forms of molecular energy are taken to be independent of each other. This means, for example, that the permitted rotational levels are independent of what vibrational level is occupied, and that the permitted rotational and vibrational levels are independent of what electronic level is occupied. This is in spite of the fact that vibrational excitation and electronic excitation both affect molecular dimensions and therefore rotational energy levels to some extent. Fortunately the error introduced by ignoring such effects is usually quite small.

Although the translational, rotational, and vibrational energies denoted, respectively, by ϵ_t, ϵ_r, and ϵ_v, are taken to be independent of each other the energies are weakly coupled, that is, there can be an exchange of energy between any two forms and an exchange of the same form of energy among the various molecules. With this provision in mind we write for the (total) energy of the *i*th level of a given molecule

$$\epsilon_i = \epsilon_{\text{tot}} = \epsilon_t + \epsilon_r + \epsilon_v \tag{4.21}$$

Now each of ϵ_t, ϵ_r, and ϵ_v may be degenerate—reference has already been made to the very high degeneracy of ϵ_t—so that, if the respective degeneracies are denoted by g_t, g_r, and g_v, a particular value of ϵ_{tot} will have a degeneracy of $g_t g_r g_v$, or

$$g_i = g_{\text{tot}} = g_t g_r g_v \tag{4.22}$$

For example, if a particular ϵ_t is given by 10^{10} quantum states, a particular ϵ_r by 10^2 quantum states, and a particular ϵ_v by 5 quantum states, the total energy, $\epsilon_t + \epsilon_r + \epsilon_v$, can result in $10^{10} \times 10^2 \times 5$ ways.

Since the partition function is still defined by (1.23), namely,

$$q = \sum_i g_i e^{-\epsilon_i/kT} \tag{4.23}$$

we can write

$$q = \sum g_t g_r g_v e^{-(\epsilon_t + \epsilon_r + \epsilon_v)/kT}$$

or

$$q = \sum (g_t e^{-\epsilon_t/kT})(g_r e^{-\epsilon_r/kT})(g_v e^{-\epsilon_v/kT}) \tag{4.24}$$

This is a sum of products. However, according to a mathematical theorem "the sum of products" equals "the product of sums," that is, we can use "the product of sums" to replace the above "sum of products." For example, $(1 - 3 + 5)(2 + 4)(3 + 7)$, which is the product of sums, can replace the sum of (all possible) products, namely, $1\cdot2\cdot3\cdot + 1\cdot2\cdot7 + 1\cdot4\cdot3 + 1\cdot4\cdot7 - 3\cdot2\cdot3 - 3\cdot2\cdot7 - 3\cdot4\cdot3 - 3\cdot4\cdot7 + 5\cdot2\cdot3 + 5\cdot2\cdot7 + 5\cdot4\cdot3 + 5\cdot4\cdot7$, both expressions being equal to 180. We rewrite (4.24), therefore, as

$$q = \sum g_t e^{-\epsilon_t/kT} \sum g_r e^{-\epsilon_r/kT} \sum g_v e^{-\epsilon_v/kT} \tag{4.25}$$

Now each of the sums on the right has the same form as (4.23), so it is appropriate to call them the partition functions for translation, rotation, and vibration, respectively. Thus we have

$$q = q_t q_r q_v \tag{4.26}$$

It may appear that we have overlooked the possibility that a given ϵ_{tot} in (4.21), in addition to arising from different quantum states for a given ϵ_t, a given ϵ_r, and a given ϵ_v, could have arisen from other possible combinations of ϵ_t, ϵ_r, and ϵ_v. Actually this possibility has not been overlooked but has been included, as will be shown in Problem 4.3.

EXAMPLE 4.3. (a) Consider a fictitious system in which there is only translational and rotational energy, with the translational energy levels being limited to 0 and 1 unit of energy (above ground) and the rotational energy levels being limited to 0 and 2 units of energy (above ground), with both kinds of energy in the same size units. Suppose, further, that both translational levels are triply degenerate, that the ground rotational level is nondegenerate, and that the other rotational level is triply degenerate. Write out the translational, rotational, and total partition functions, term by term.

(b) Show that the product of the first two equals the third.

ANSWER. (a) $q_t = 3 + 3e^{-1/kT}$; $q_r = 1 + 3e^{-2/kT}$. With two possible values for ϵ_t and two for ϵ_r, ϵ_i can have four values: $0 + 0 = 0$, $1 + 0 = 1$, $0 + 2 = 2$, and $1 + 2 = 3$ units, with respective degeneracies of $3 \times 1 = 3$, $3 \times 1 = 3$, $3 \times 3 = 9$, $3 \times 3 = 9$. Therefore

$$q_{\text{tot}} = 3 + 3e^{-1/kT} + 9e^{-2/kT} + 9e^{-3/kT}$$

(b)

$$q_t q_r = (3 + 3e^{-1/kT})(1 + 3e^{-2/kT})$$

$$= 3 + 3e^{-1/kT} + 9e^{-2/kT} + 9e^{-3/kT} = q_{\text{tot}}$$

as found in (a).

With the many kinds of energy, ϵ_t, ϵ_r, ϵ_v, etc. and ϵ_{tot} for every particle, it is appropriate to enquire to what extent the Boltzmann distribution applies to them, collectively and individually. It should be clear from Section 4.1 that it does apply to ϵ_{tot} which is represented by ϵ_i in (4.5) and (4.6). But does it apply individually to ϵ_t, ϵ_r, and ϵ_v? The answer is "yes." To show this, consider n_i/N for the particular set of values of ϵ_t^*, ϵ_r^*, and ϵ_v^*. According to (4.6) and (4.26)

$$\frac{n_i}{N} = \frac{g_t g_r g_v e^{-(\epsilon^*_t + \epsilon^*_r + \epsilon^*_v)/kT}}{q_{\text{tot}}} = \frac{g_t e^{-\epsilon^*_t/kT}}{q_t} \frac{g_r e^{-\epsilon^*_r/kT}}{q_r} \frac{g_v e^{-\epsilon^*_v/kT}}{q_v} \qquad (4.27)$$

which gives the fraction of all the molecules with this particular set. Similarly, the fraction with the particular set $\epsilon_t^*, \epsilon_r^*, \epsilon_v^{**}$ will be

$$\frac{g_t e^{-\epsilon^* t/kT}}{q_t} \frac{g_r e^{-\epsilon^* r/kT}}{q_r} \frac{g_v e^{\epsilon^{**} v/kT}}{q_v}$$

and the fraction with the particular set $\epsilon_t^*, \epsilon_r^*, \epsilon_v^{***}$ will be

$$\frac{g_t e^{-\epsilon^* t/kT}}{q_t} \frac{g_r e^{-\epsilon^* r/kT}}{q_r} \frac{g_v e^{-\epsilon^{***} v/kT}}{q_v}$$

and so on. Writing this fraction for all possible ϵ_v's and adding the fractions will give the fraction of all the molecules with the particular values ϵ_t, ϵ_r, and *any* value of ϵ_v, or

$$\frac{g_t e^{-\epsilon^* t/kT}}{q_t} \frac{g_r e^{-\epsilon^* r/kT}}{q_r} \times 1$$

In a similar fashion we could write expressions for n_i/N for the particular sets $\epsilon_t^*, \epsilon_r^{**}$, and any value of ϵ_v; $\epsilon_t^*, \epsilon_r^{***}$, and any value of ϵ_v; . . . ; and show that, for the value ϵ_t^* and any value of ϵ_r and any value of ϵ_v, the fraction is

$$\frac{g_t e^{-\epsilon^* t/kT}}{q_t} \times 1 \times 1$$

which may be written

$$n_t^* = \frac{N}{q_t} g_t e^{-\epsilon^* t/kT} \tag{4.28}$$

Thus it turns out that the Boltzmann distribution law applies equally well to the translational energies. An analogous argument shows, of course, that it applies to the rotational and to the vibrational energies, so that

$$n_r^* = \frac{N}{q_r} g_r e^{-\epsilon^*_r/kT} \tag{4.29}$$

and

$$n_v^* = \frac{N}{q_v} g_v e^{-\epsilon^*_v/kT} \tag{4.30}$$

Equations (4.28), (4.29), and (4.30), each divided by N, give the probability of a particle having translational energy ϵ_t^*, rotational energy ϵ_r^*, and vibrational energy ϵ_v^*, respectively. The product of these probabilites is the probability of its having all three values simultaneously, which brings us back to (4.27).

EXAMPLE 4.4. (a) For a certain system consisting of an ideal gas, $q_t = 10^{30}$, $q_r = 10^2$, and $q_v = 1.10$ at 300 K and 1 atm. What fraction of the molecules has a translational energy of 6.0×10^{-21} J for which $g_t = 10^5$? What fraction has a rotational energy of 4.0×10^{-21} J for which $g_r = 30$? What fraction has a vibrational energy of 1.00×10^{-21} J for which $g_v = 1$?

(b) What fraction of the molecules has a total energy equal to the sum of the energies given in (a), namely, 11.0×10^{-21} J?

ANSWER. (a) From (4.28) $n_t^*/N = 10^5 \times \exp(-6.0 \times 10^{-21}/300k)/10^{30} = 2.35 \times 10^{-26}$. Analogously $n_r^*/N = 0.114$ from (4.29) and $n_v^*/N = 0.714$ from (4.30).

(b) $(n_t^*/N)(n_r^*/N)(n_v^*/N) = 1.91 \times 10^{-27}$.

4.4. BREAKDOWN OF CONTRIBUTIONS TO THERMODYNAMIC FUNCTIONS

The factorization of q shown in the previous section makes it possible to find out what part of the value for a given thermodynamic function such as E or S is the result of its translational energy, what part is the result of its rotational energy, and so on, and consequently throws light on the interpretation of the data.

If one examines (4.9), (4.11), (4.12), (4.14), (4.17), (4.19), and (4.20) it is apparent that it is always the logarithm of q, not q itself which appears in the expressions for the thermodynamic functions. By taking the logarithm of (4.26) prior to using it in the above-mentioned equations,

$$\ln q = \ln q_t + \ln q_r + \ln q_v \tag{4.31}$$

and it can be seen that every thermodynamic function can be expressed in terms of the contributions of its several parts. With the energy E, for example, combination of (4.31) and (4.9) makes it possible to write

$$E = NkT^2 \left(\frac{\partial \ln q_t}{\partial T} \right)_V + NkT^2 \frac{d \ln q_r}{dT} + NkT^2 \frac{d \ln q_v}{dT} \tag{4.32}$$

where the partial notation has been dropped from the last two terms, since q_r and q_v are independent of volume even though q_t is not, as will be seen in the next chapter. It is clear, then, that we can identify the first term on the right with the translational energy, the second with the rotational energy, and so on. Therefore

$$E_t = NkT^2 \left(\frac{\partial \ln q_t}{\partial T} \right)_V \tag{4.33}$$

$$E_r = NkT^2 \frac{d \ln q_r}{dT} \tag{4.34}$$

$$E_v = NkT^2 \frac{d \ln q_v}{dT} \tag{4.35}$$

The extension of this breakdown process to C_V, H, and C_P is obvious.

With S, A, and G, however, additional comment is required because $\ln(q/N)$, rather than $\ln q$ is involved. Combining (4.31) and (4.17), and using (4.32)–(4.35) yields

$$S = kN \ln \frac{q_t q_r q_v}{N} + \frac{E_t + E_r + E_v}{T} + kN \tag{4.36}$$

and we are faced with deciding where to assign the N in the denominator of the first term, and where to assign the kN term. The correct choice is to include these with the translation term, namely,

$$S = kN \ln \frac{q_t}{N} + \frac{E_t}{T} + kN + kN \ln q_r + \frac{E_r}{T} + kN \ln q_v + \frac{E_v}{T} \tag{4.37}$$

because they arose from the indistinguishability of the molecules, and the latter was precisely the result of their translational motion. It is thus possible to break down the various contributions to S as follows:

$$S_t = kN \ln \frac{q_t}{N} + \frac{E_t}{T} + kN \tag{4.38}$$

$$S_r = kN \ln q_r + \frac{E_r}{T} \tag{4.39}$$

$$S_v = kN \ln q_v + \frac{E_v}{T} \tag{4.40}$$

Moreover, for the free energies, using (4.19) and (4.20),

$$A_t = -kNT \ln \frac{q_t}{N} - kNT, \qquad A_r = -kNT \ln q_r, \qquad A_v = -kNT \ln q_v \tag{4.41}$$

and

$$G_t = -kNT \ln \frac{q_t}{N}, \qquad G_r = -kNT \ln q_r,$$
$$G_v = -kNT \ln q_v \quad (4.42)$$

EXAMPLE 4.5. (a) For certain diatomic ideal gas at 300 K and 1 atm $q_t = 10^{30}$, $q_r = 10^2$, and $q_v = 1.000$. Under these conditions its translational and rotational energies have the classical values, $\frac{3}{2}RT$ and RT, respectively. What is the vibrational energy?

(b) Find the molar entropy.

ANSWER. (a) Since q_v is unity only the ground vibrational state is populated, so there is no vibrational energy (above ground).

(b)

$$E_t = \tfrac{3}{2}R(300) = 3740 \text{ J/mol}$$
$$E_r = R(300) = 2493 \text{ J/mol}$$

Substitution in (4.38) and (4.39) gives

$$S_t = R \ln \frac{10^{30}}{L} + \frac{3740}{300} + R = 140 \text{ J/K mol}$$
$$S_r = R \ln 10^2 + \frac{2493}{300} = 47 \text{ J/K mol.}$$

Note: Adding the two results in (b), or using (4.18) directly, gives $S = 187$ J/K mol. (There is no vibrational contribution.)

We have thus succeeded in breaking down the thermodynamic functions into the contributions of the various kinds of energy, each contribution being expressed in terms of its own partition function. There remains to show how these partition functions are actually determined, and this is the purpose of the following two chapters. We will then have at our disposal all the tools needed for the calculation of the thermodynamic functions of ideal gases.

4.5. PROOF THAT $\beta = 1/kT$

In the derivation of the Boltzmann distribution law in Section 1.2 it was stated without proof that β, one of the constants used in the method of undetermined multipliers, equals $1/kT$, although it was known that it had the dimensions of energy^{-1}. We now undertake to prove that $\beta = 1/kT$.

Consider a closed system of distinguishable molecules in a vessel of fixed volume. Both N and V are therefore constants. We apply the following four relationships, already developed, to such a system:

(1) the expression for entropy in terms of q and E,

$$S = kN \ln q + k\beta E \tag{3.12}$$

(2) the definition of partition function

$$q = \sum_i g_i e^{-\beta\epsilon_i} \tag{1.23}$$

(3) the expression for the energy of a system, E, in terms of q,

$$E = \frac{N}{q} \sum_i \epsilon_i g_i e^{-\beta\epsilon_i} \tag{3.1}$$

(4) the differential equation from classical thermodynamics for a closed system subject to flow of energy into or out of the system at constant volume,

$$\left(\frac{\partial S}{\partial E}\right)_V = \frac{1}{T} \tag{2.12}$$

In the first three of these kT has been replaced by $1/\beta$ since this equality has yet to be shown. Differentiation of (3.12) gives

$$\left(\frac{\partial S}{\partial E}\right)_V = \left(\frac{kN}{q}\right)\left(\frac{\partial q}{\partial E}\right)_V + k\beta + kE\left(\frac{\partial \beta}{\partial E}\right)_V$$

since both β and E depend on E. We rewrite this as

$$\left(\frac{\partial S}{\partial E}\right)_V = \left(\frac{kN}{q}\right)\left(\frac{\partial q}{\partial \beta}\right)_V\left(\frac{\partial \beta}{\partial E}\right)_V + k\beta + kE\left(\frac{\partial \beta}{\partial E}\right)_V \tag{4.43}$$

and proceed to evaluate $(\partial q/\partial \beta)_V$ from (1.23), namely,

$$\left(\frac{\partial q}{\partial \beta}\right)_V = -\sum_i \epsilon_i g_i e^{-\beta\epsilon_i}$$

Substitution of this, and of (3.1) into (4.43) yields

$$\left(\frac{\partial S}{\partial E}\right)_V = \frac{kN}{q}\left(-\sum_i \epsilon_i g_i e^{-\beta\epsilon_i}\right)\left(\frac{\partial \beta}{\partial E}\right)_V + k\beta + k\frac{N}{q}\left(\sum_i \epsilon_i g_i e^{-\beta\epsilon_i}\right)\left(\frac{\partial \beta}{\partial E}\right)_V$$

which reduces immediately to $k\beta$. Finally, application of (2.12) gives

$$\beta = \frac{1}{kT} \tag{4.44}$$

Had we started with a system of indistinguishable molecules (3.12) would have been replaced by (4.17), namely, $S = kN \ln (q/N) + k\beta E + kN$, but the final result would have been the same.

PROBLEMS

4.1. Show that (4.2) and (4.3) both reduce to (4.4) for $g_i \gg n_i$.

***4.2.** Molecules of argon gas possess only translational energy. Following the procedure described in the next chapter, its partition function, when 1 mole of it occupies a volume of 2.4465×10^{-2} m^3 at 298.15 K and 1 atm, can be shown to be 5.9762×10^{30}. Its energy has the classical value, $\frac{3}{2}RT$, or 3718.4 J/mol. Find its molar entropy under these conditions.

4.3. (a) The molecules of a hypothetical system have only two kinds of energy, translational and rotational. The only permitted translational levels are $\epsilon_{t0} = 0$, $\epsilon_{t1} = 1$, and $\epsilon_{t2} = 2$ energy units, with $g_{t0} = 1$, $g_{t1} = 2$, and $g_{t2} = 3$. The only rotational levels are $\epsilon_{r0} = 0$ and $\epsilon_{r1} = 2$ energy units, with $g_{r0} = 1$ and $g_{r1} = 2$. The energy units for translation are the same size as those for rotation. List the possible values of ϵ_{tot} which are available, and indicate what values of the translational and rotational energies give rise to each.

(b) What characteristic does this example possess which was lacking in Example 4.3?

(c) Write out the expressions for q_t and q_r, term by term, and evaluate $q_t q_r$. Does this agree with q_{tot} expressed in terms of ϵ_{tot}?

4.4. It will be shown in Chapter 5 that q_t is directly proportional to the volume, V. Examine (4.38) to determine what will be the effect of doubling V (without changing the temperature) on the entropy of a mole of ideal gas.

***4.5.** Find expressions for the translational, rotational, and vibrational contributions to the enthalpy of a mole of ideal gas in terms of q_t, q_r, and q_v.

Chapter Five

THERMODYNAMIC FUNCTIONS FOR IDEAL GASES—PART I

5.1. THE TRANSLATIONAL PARTITION FUNCTION

In the computation of partition functions one relies heavily upon the results of quantum mechanics for the permitted energies. As stated in (4.1) the permitted translational energies for a molecule of ideal gas of mass m in a cubic container of edge a are given by

$$\epsilon_i = (n_x^2 + n_y^2 + n_z^2)h^2/8ma^2 \tag{5.1}$$

where the quantum numbers can have integral values from unity upwards. The translational partition function, therefore, is given by

$$q_t = \sum_{n_x} \sum_{n_y} \sum_{n_z} \exp[-(n_x^2 + n_y^2 + n_z^2)h^2/8ma^2kT] \tag{5.2}$$

where $\Sigma\ \Sigma\ \Sigma$ means that all possible combinations of n_x, n_y, and n_z must be included in the summation. In accordance with the alternative description of partition function mentioned at the end of Section 1.4 we have summed over all the quantum states and omitted the degeneracies, recognizing that many of the terms in the resulting series will be identical. Equation

(5.2) can, of course, be rewritten as

$$q_t = \sum_{nx} \sum_{ny} \sum_{nz} \exp(-n_x^2 h^2/8ma^2kT) \exp(-n_y^2 h^2/8ma^2kT) \exp(-n_z^2 h^2/8ma^2kT) \quad (5.3)$$

This "sum of products" can be replaced by the "product of sums," as was done in Section 4.3, to give

$$q_t = \sum_{nx} \exp(-n_x^2 h^2/8ma^2kT) \sum_{ny} \exp(-n_y^2 h^2/8ma^2kT) \sum_{nz} \exp(-n_z^2 h^2/8ma^2kT) \quad (5.4)$$

Since the numerical values of all three factors on the right are the same we now write

$$q_t = \left[\sum_n \exp(-n^2 h^2/8ma^2kT) \right]^3 \quad (5.5)$$

Because the difference between each of the terms in (5.5) and the one following it is so small the expression can be replaced by

$$q_t = \left[\int_1^\infty \exp(-n^2 h^2/8ma^2kT)\, dn \right]^3 \quad (5.6)$$

which differs negligibly from

$$q_t = \left[\int_0^\infty \exp(-n^2 h^2/8ma^2kT)\, dn \right]^3 \quad (5.7)$$

If $h^2/8ma^2kT$, which is a constant for a given molecular species at a given temperature, is replaced temporarily by c we have

$$q_t = \left[\int_0^\infty e^{-cn^2}\, dn \right]^3$$

and this is the cube of a standard integral the value of which is $\frac{1}{2}(\pi/c)^{1/2}$. We may therefore write

$$q_t = \left[\frac{1}{2} \left(\frac{\pi}{c} \right)^{1/2} \right]^3 = \frac{(2\pi mkT)^{3/2} a^3}{h^3}$$

Finally, since a^3 is the volume of the cube, and since the shape of the container has no effect on the properties of the gas,

$$q_t = \frac{(2\pi mkT)^{3/2} V}{h^3} \tag{5.8}$$

which is the desired expression. This translational partition function is a property of the molecule of mass m at temperature T moving in a volume V. Its value does not depend on the quantity of material present unless this affects the volume occupied. It is dimensionless, as are all partition functions and, under ordinary conditions of temperature and pressure, has an order of magnitude of about 10^{30}. Note that q_t is directly proportional to V as mentioned in Problem 4.4.

EXAMPLE 5.1. Evaluate q_t for $O_2(g)$ molecules at 273.15 K occupying a volume of 22.414 dm^3. The molecular weight is 31.99 g/mol.

ANSWER. Since $m = 31.99 \times 10^{-3}/L = 5.312 \times 10^{-26}$ kg and $V = 22.414 \times 10^{-3}$ m^3, $q_t = [(2\pi)(5.312 \times 10^{-26})(1.3807 \times 10^{-23})(273.15)]^{3/2} \times (22.414 \times 10^{-3})/6.626 \times 10^{-34})^3 = 3.441 \times 10^{30}$.

Having evaluated q_t the obvious thing to do now is to see whether it leads to known experimental values for the state functions such as E, C_p, and S. We must confine ourselves to monatomic gases because only for these can we expect there to be no contributions from rotation and vibration, the partition functions we have yet to consider. Using (4.33)

$$E_t = kNT^2\left[\frac{\partial}{\partial T}\ln\frac{(2\pi mkT)^{3/2}V}{h^3}\right]_V$$
$$= kNT^2\frac{d}{dT}[\ln T^{3/2} + \ln(\text{const})]$$
$$= kNT^2(T^{-3/2})(3/2)(T^{1/2})$$

or

$$E_t = \tfrac{3}{2}kNT \tag{5.9}$$

Therefore

$$E_t\text{ (per mole)} = \tfrac{3}{2}RT \tag{5.10}$$

and

$$C_{V_t}\text{ (per mole)} = \tfrac{3}{2}R \tag{5.11}$$

which are identical to the classical results. It follows, of course, that since, for one mole, $PV = RT$,

$$H_t\text{ (per mole)} = \tfrac{3}{2}RT + RT = \tfrac{5}{2}RT \tag{5.12}$$

and

$$C_{P_t}\text{ (per mole)} = \tfrac{5}{2}R \tag{5.13}$$

all of which are amply supported by experiment.

To find the translational entropy we substitute (5.8) and (5.9) into (4.38) giving

$$S_t = kN \ln \frac{(2\pi mkT)^{3/2}V}{Nh^3} + \frac{5}{2} kN \qquad (5.14)$$

which is known as the Sackur–Tetrode equation. It is convenient to eliminate V by writing it as a function of T and P, namely, $V = RT/P$ (per mole), being careful to express the quantities in consistent units. For example, using the SI system m must be in kg, R in J/K mol, T in K, V in m^3, k in J/K, h in J s and P in N/m^2. [1 Newton (N) = 1 kg m^{-1} s^{-2} and 1 atm = 101,325 N/m^2.] For the standard pressure of 1 atm, therefore, (5.14) reduces to

$$S_t^0 \text{ (per mole)} = R \ln \frac{(2\pi mkT)^{3/2}RT}{101{,}325Lh^3} + \frac{5}{2} R$$

The superscript zero signifies standard pressure. This can be further simplified. Since, in the SI system, m must be in kg, it can be replaced by $M/1000\ L$, where M is the usual molecular weight *in g/mol*, giving

$$S_t^0 \text{ (per mole)} = R \ln M^{3/2}T^{5/2} + R \ln [(2\pi k/1000L)^{3/2}(k/101{,}325h^3)] + \tfrac{5}{2}R$$

which, on substitution for the constants, reduces to

$$S_t^0(\text{J/K mol}) = \tfrac{3}{2}R \ln M + \tfrac{5}{2}R \ln T - 9.685 \qquad (5.15)$$

For the common temperature of 298.15 K (5.15) becomes

$$S_{t_{298}}^0 \text{ (J/K mol)} = \tfrac{3}{2}R \ln M + 108.745 \qquad (5.16)$$

EXAMPLE 5.2. Calculate the standard entropy of Ar(g). $M = 29.948$ g/mol at 298.15 K.

ANSWER. Since Ar is monatomic, $S^0 = S_t^0$. Using (5.16), $S_{298}^0 = \frac{3}{2}(8.3144) \ln 39.948 + 108.745 = 154.735$ J/K mol. (The third law value, corrected to ideal gas behavior, is 154.6, the small difference being within experimental error in the third law result.)

It is seen from (5.15) that, for a given pressure, the contribution to the molar entropy of the translational motion depends only on M and T. This is reasonable since the only properties which can characterize an independent, moving point mass in a given volume (the molar volume) are its mass and its velocity. The constant term in this equation depends on the pressure. If we write the equation for a given gas at two temperatures but the same pressure, and subtract one equation from the other, we find that, for an isobaric change in temperature from T_1 to T_2, the change in the molar entropy of a monatomic ideal gas is

$$\Delta S = \frac{5}{2} R \ln \frac{T_2}{T_1} \tag{5.17}$$

which agrees with the classical result, $\Delta S = C_P \ln (T_2/T_1)$, through (5.13).

We may, of course, also express A_t and G_t in terms of M, T, and P through (4.41) and (4.42), but this will be left to the reader. It is, perhaps, worthwhile to point out that even though q_t, S_t, A_t, and G_t all depend on V (and therefore on P), E_t, C_V, and C_P do not depend on it.

5.2. THE MECHANICS OF ROTATION

As a preliminary to considering the contribution of rotational energy (tumbling motion) of the molecules to the thermodynamic functions it is necessary to understand something of the classical mechanics of rotation. Just as the kinetic energy of a moving molecule is proportional to its mass so the rotational energy is proportional to its moment of inertia, so we shall examine the meaning of this term and how it is computed.

A monatomic gas molecule, with practically all its mass concentrated in the tiny nucleus, has no detectable tumbling motion. For example, we have seen that by considering only the translational contribution for Ar(g) in Example 5.2 the experimental entropy is reproduced, and the same is true for other monatomic gases. Molecules with two or more atoms, however, do possess rotational energy, and this constitutes a sizable contribution which can be as large as, but is usually smaller than, the translational contribution.

For the present purpose we regard a molecule as a rigid array of atoms in space, each represented by a point mass. Although it is not essential to do so, let us see how the center of mass of the molecule is located. The latter can be thought of as that point within the molecule (sometimes coinciding with one of the atoms) at which, if the molecule were suspended at that point, it would show no tendency to rotate in any direction. It is analogous to locating the point, *P*, on a weightless beam (AB of Figure 5.1) from the ends of which are hung weights of different masses, such that suspension of the beam at that point results in no rotation. Clearly, *P* must

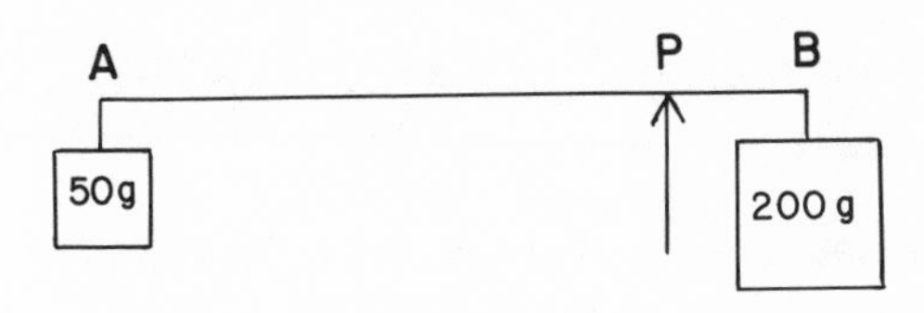

Figure 5.1. Illustration of center mass.

be located such that $AP = 4 \times PB$ if the mass at B is four times that at A.

To find the position of P in a molecule, which, in general, will be an array of point masses in three dimensions, some arbitrary but reasonable location is chosen as an origin of a rectangular system of coordinate axes, and the coordinates of every atom with respect to these axes is determined from a knowledge of the bond lengths and bond angles. (If the molecule has a plane of symmetry the origin should be placed somewhere on it—if there is more than one plane of symmetry it should be placed on the line of intersection of the planes. The orientation of the axes about this origin should, if possible, be such that one of them coincides with an axis of symmetry or that two of the coordinate axes lie in a plane of symmetry.) If each atom of mass m_i is thus found to have coordinates x_i, y_i, z_i, and if x', y', z', are the coordinates of the center of mass with respect to the same axes, then the solution of the following equations for x', y', and z' locates the center of mass, P:

$$\sum_i m_i(x_i - x') = 0, \qquad \sum_i m_i(y_i - y') = 0, \qquad \sum_i m_i(z_i - z') = 0 \qquad (5.18)$$

With x', y', and z' now known, the coordinates of P have been established. If the origin is now transferred to P the *new* coordinates will satisfy the equations

$$\sum_i m_i x_i = 0, \qquad \sum_i m_i y_i = 0, \qquad \sum_i m_i z_i = 0 \qquad (5.19)$$

which is the mathematical criterion for the center of mass. Of course if the molecule has more than one axis of symmetry the position of its center of mass is at their intersection.

It will be evident that the presence of symmetry facilitates finding the center of mass. In this event the formal procedure described above can be greatly abbreviated, as will be clear from the following examples.

Our first illustration will be a very simple one, the HD molecule, for which the atomic weights will be taken as H = 1.00 and D = 2.00 g/mol, and the bond length 0.75 Å or 75 pm. Since P, the center of mass, must be somewhere on the line joining H to D (true for all diatomic molecules), we can choose an origin at H (Figure 5.2) giving the coordinates $x = 0$ for H and $x = 75$ for D. Using only the first of Eqs. (5.18)—since the molecule is only one-dimensional—we have $(1.00/L)(0 - x') + (2.00/L)(75 - x') = 0$, or $x' = 50$ pm. Thus the center of mass is 50 pm from the H atom and 25 pm from the D.

For a second illustration let us determine the center of mass for the H_2O molecule, having given the atomic weights as H = 1.00 and O = 16.0 g/mol, the O–H bond length as 96 pm, and the H–O–H bond angle as 105°. The molecule is planar and has a twofold axis of symmetry lying in this plane and bisecting the bond angle. Thus the center of mass, P, will lie in the plane of the molecule and somewhere on the bisector as in Figure 5.3. If the oxygen atom is chosen as the origin of a set of coordinates, with the y axis coinciding with the bisector of the bond angle, it is evident that only the y coordinate of P need be found. Since OQ = 96 cos 52.5 = 58.4, the y coordinates of the atoms are H(1) = 58.4; O = 0; H(2) = 58.4. Applying the second of Eqs. (5.18) gives $(1.00/L)(58.4$

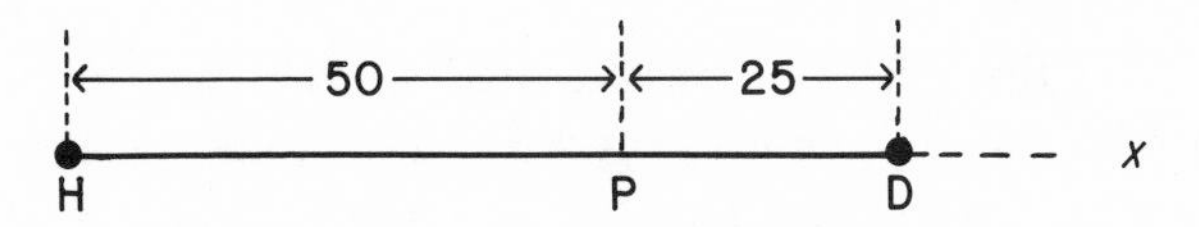

Figure 5.2. Center of mass of HD molecule.

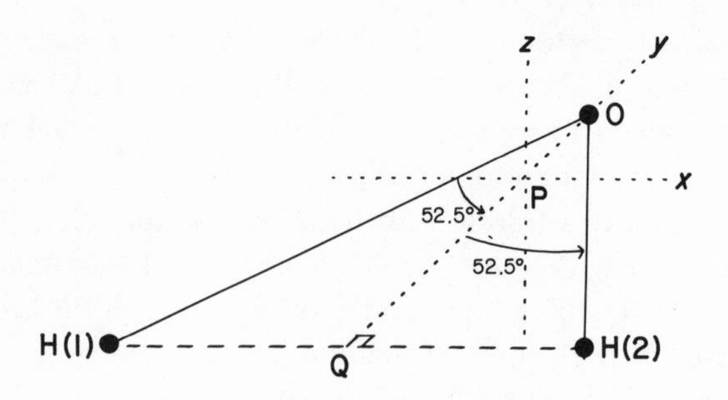

Figure 5.3. Center of mass of H_2O molecule.

$- y') + (16.0/L)(0 - y') + (1.00/L)(58.4 - y') = 0$, the solution of which is $y' = 6.5$. P is therefore 6.5 pm from the oxygen atom; it is close to the latter because most of the mass of the molecule is concentrated in the oxygen.

EXAMPLE 5.3. From symmetry considerations estimate the location of the center of mass of the following molecules: N_2, CO, C_6H_6 (benzene), CO_2 (linear), O_3 (angular), HOD (angular).

ANSWER. N_2: P will lie midway between the atoms. CO: P will lie between the two atoms but slightly nearer to the oxygen. C_6H_6: Since it is planar and has a sixfold symmetry axis perpendicular to this plane P will lie at the center of the hexagon of C atoms. CO_2: Since C lies midway between the two O's it will coincide with P. O_3: P will lie in the plane of the molecule on the bisector of the O–O–O angle, and within the triangle formed by the three atoms. HOD: P will lie in the plane of the molecule, within the triangle formed by the three atoms, but on the side of the angular bisector towards the D atom.

With the center of mass located we can now proceed to determine the moments of inertia, I, of a molecule. There is a moment of inertia about *any* axis through P, and it is defined by

$$I = \sum_i m_i r_i^2 \tag{5.20}$$

where m_i is the atomic mass of the ith atom and r_i is the *perpendicular* distance of that atom from the chosen axis. We can, of course, set up a rectangular coordinate system with the origin at P and, *regardless* of how the axes are oriented, speak of a moment of inertia about any one of the three axes. Each atom will have its own x_i, y_i, z_i values with respect to these axes, and the moments of inertia about the x, y, and z axes, denoted by I_{xx}, I_{yy}, and I_{zz}, will, according to (5.20), be given by

$$I_{xx} = \sum_i m_i(y_i^2 + z_i^2) \tag{5.21}$$

$$I_{yy} = \sum_i m_i(x_i^2 + z_i^2) \tag{5.22}$$

$$I_{zz} = \sum_i m_i(x_i^2 + y_i^2) \tag{5.23}$$

There are also three "products of inertia," defined as follows:

$$I_{xy} \quad \text{or} \quad I_{yx} = \sum_i m_i x_i y_i \tag{5.24}$$

$$I_{xz} \quad \text{or} \quad I_{zx} = \sum_i m_i x_i z_i \tag{5.25}$$

$$I_{yz} \quad \text{or} \quad I_{zy} = \sum_i m_i y_i z_i \tag{5.26}$$

Notice that the moments of inertia will always be positive but the products of inertia may be negative.

If we were to calculate the moments of inertia for all possible orientations of the axes (with the origin remaining fixed at the center of mass) we would find that there is a particular orientation for which one of the I's has a maximum value, another has a minimum value, and the third is either equal to one of the others or intermediate in value between the other two. When the axes are in that particular orientation they are said to correspond to the principal axes of the molecule, and the three corresponding moments are called the principal moments of inertia, denoted by I_A, I_B, and I_C, respectively. Moreover, in that particular orientation the products of inertia all reduce to zero. It is usual to label the axes such that $I_A < I_B < I_C$.

In the general case all three principal moments have different values but, depending on the molecular symmetry, the cases shown in Table 5.1 arise with the indicated names. A planar molecule can be either a symmetric or an asymmetric top. For monatomic molecules $I_A = I_B = I_C = 0$. A symmetric top must have an axis of symmetry which is at least threefold.

Most molecules have no symmetry. The location of the principal axes for such a molecule is somewhat involved. The presence of any symmetry makes the calculation easier. An axis of symmetry, for example, in addition to passing through the center of mass, must itself be one of the principal axes.

For linear molecules, such as N_2, HBr, CO_2, $BeCl_2$, and N_2O, there are two equal, nonzero, principal moments of inertia. They are the moments about the two axes at right angles to the axis of the molecule and at right angles to each other. As they are equal, it is common to say that a linear molecule has one moment of inertia, given the symbol I. For *diatomic* molecules the calculation of I is particularly simple. If the H

Table 5.1. Categories of Rotating Molecules

Characteristics	Name	Example
$I_A = I_B = I_C$	Spherical top	CCl_4
$I_A = I_B \neq I_C$	Symmetric top	NH_3
$I_A \neq I_B \neq I_C$	Asymmetric top	CH_2Cl_2
$0 = I_A \neq I_B = I_C = I$	Linear	$HC{\equiv}CH$
$I_A + I_B = I_C$	Planar	C_6H_6

and D of Figure 5.2 are replaced by "1" and "2", respectively, and if r_1 and r_2 are their respective distances from P, then $r = r_1 + r_2$ is the bond length and $I = m_1r_1^2 + m_2r_2^2$ according to (5.20). However, according to (5.19), $m_1r_1 = m_2r_2$, from which $r_1 = m_2r_2/m_1$, so $r = r_1 + r_2 = r_2(m_1 + m_2)/m_1$, from which $r_2 = m_1r/(m_1 + m_2)$. Similarly, $r_1 = m_2r/(m_1 + m_2)$. Substituting these expressions for r_1 and r_2 in (5.20) gives

$$I = \frac{m_1m_2}{m_1 + m_2} r^2 \tag{5.27}$$

which is usually abbreviated

$$I = \mu r^2 \tag{5.28}$$

where μ, called the reduced mass, stands for $m_1m_2/(m_1 + m_2)$.

EXAMPLE 5.4. Compute the moment of inertia for the HD molecule given that the atomic weights are H = 1.0 and D = 2.0 g/mol, and that the bond length is 75 pm.

ANSWER. $\mu = (1.0/L)(2.0/L)(10^{-3})/(1.0/L + 2.0/L) = 1.1 \times 10^{-27}$ kg. Therefore, by (5.28) $I = 1.1 \times 10^{-27}(75^2) = 6.2 \times 10^{-24}$ kg pm^2 or 6.2×10^{-48} kg m^2.

It will be noticed that the method just described for finding I for a diatomic molecule does not require locating the center of mass. For linear *poly*atomic molecules (for which the above procedure is not appropriate) one can do this as follows: If the molecular axis is called the x axis and any point on it is chosen as origin (preferably one of the atoms), and if x_i is the coordinate of each atom with respect to it, then

$$I = \sum_i m_ix_i^2 - (1/M)\left(\sum_i m_ix_i\right)^2 \tag{5.29}$$

where the m_i's and the molecular weight, M, are in the same units, conveniently g/mol. Alternatively, one may use the relation

$$I = (1/M)(\tfrac{1}{2}) \sum_i \sum_j m_i m_j r_{ij}^2 \qquad (5.30)$$

in which r_{ij} is the distance apart of the ith and jth atoms, and the terms in the sum are obtained by pairing *every* atom with *every other* atom, to give n^2 terms altogether for a molecule with n atoms.

EXAMPLE 5.5. Find I for NNO (linear) using (5.29) if the bond lengths are N–N = 113, N–O = 119 pm and the atomic weights are N = 14.0, O = 16.0 g/mol.

ANSWER. If the end nitrogen is chosen for origin the coordinates are N(0), N(113), O(232). Since M = 44.0 g/mol, $I = 14.0(0^2) + 14.0(113^2) + 16.0(232^2) - (1/44.0)[14.0(0) + 14.0(113) + 16.0(232)]^2 = 4.03 \times 10^5$ amu pm^2 or $4.03 \times 10^5/L = 6.69 \times 10^{-19}$ g pm^2 $= 6.69 \times 10^{-46}$ kg m^2.

For nonlinear molecules the computation of the moments of inertia is more complicated. However, we shall see that it is not the individual moments of inertia which are required for the calculation of thermodynamic properties but their product, $I_A I_B I_C$. The latter can be readily evaluated without determining the individual values or the orientation of the principal axes by means of the relation

$$I_A I_B I_C = \begin{vmatrix} I_{xx} & -I_{xy} & -I_{xz} \\ -I_{yx} & I_{yy} & -I_{yz} \\ -I_{zx} & -I_{zy} & I_{zz} \end{vmatrix} \qquad (5.31)$$

One procedure is to locate the center of mass as already described, set up *any* convenient rectangular system of axes with the origin at the center of mass, find the coordinates of the atoms, evaluate I_{xx}, I_{yy}, I_{zz}, I_{xy}, I_{xz}, and I_{yz}, and use (5.31) to compute $I_A I_B I_C$. However, it is actually not necessary to locate the center of mass if one wishes merely to evaluate $I_A I_B I_C$. One may set up *any* convenient rectangular coordinate system with *any* origin (such as one on a symmetry axis or symmetry plane or both), determine the coordinates (x_i, y_i, z_i) of every atom with respect to these axes, and then evaluate A, B, C, D, E, and F according to the following definitions, where the m_i's and M are, for convenience, in g/mol and the coordinates in pm:

$$A = \sum_i m_i(y_i^2 + z_i^2) - (1/M)\left(\sum_i m_i y_i\right)^2 - (1/M)\left(\sum_i m_i z_i\right)^2$$

$$B = \sum_i m_i(x_i^2 + z_i^2) - (1/M)\left(\sum_i m_i x_i\right)^2 - (1/M)\left(\sum_i m_i z_i\right)^2$$

$$C = \sum_i m_i(x_i^2 + y_i^2) - (1/M)\left(\sum_i m_i x_i\right)^2 - (1/M)\left(\sum_i m_i y_i\right)^2$$

$$D = \sum_i m_i x_i y_i - (1/M)\left(\sum_i m_i x_i\right)\left(\sum_i m_i y_i\right)$$

$$E = \sum_i m_i x_i z_i - (1/M)\left(\sum_i m_i x_i\right)\left(\sum_i m_i z_i\right)$$

$$F = \sum_i m_i y_i z_i - (1/M)\left(\sum_i m_i y_i\right)\left(\sum_i m_i z_i\right)$$

Finally, $I_A I_B I_C$ is found by evaluating the following determinant:

$$I_A I_B I_C = \begin{vmatrix} A & D & -E \\ -D & B & -F \\ -E & -F & C \end{vmatrix} \tag{5.32}$$

The result, in amu^3 pm^6, may be converted to kg^3 m^6 by multiplying by $10^{-81}/L^3$ or by 4.5790×10^{-153}. This will be illustrated in Problem 5.7.

5.3. THE ROTATIONAL PARTITION FUNCTION FOR LINEAR MOLECULES

As mentioned earlier, we shall assume what is not always true, namely, that the dimensions of a molecule remain the same regardless of the speed of rotation, that is, we assume it to be a rigid rotor. Its moments of inertia will therefore be regarded as constant. (The results can be corrected where this assumption is not warranted.)

For linear molecules the permitted rotational energies are given, according to quantum mechanics, by

$$\epsilon_J = \frac{J(J + 1)h^2}{8\pi^2 I} \tag{5.33}$$

where J, the rotational quantum number, can have integral values from zero to infinity. The degeneracy of the Jth level is $2J + 1$, so that the rotational partition function is

$$q_r = \sum_J (2J + 1) \exp[-J(J + 1)h^2/8\pi^2 IkT] \tag{5.34}$$

It is convenient to make the substitution

$$\theta_r = h^2/8\pi^2 Ik \tag{5.35}$$

where θ_r is called the characteristic rotational temperature, so that

$$q_r = \sum_J (2J + 1) \exp[-J(J + 1)\theta_r/T] \tag{5.36}$$

Clearly, θ_r depends only on I. For nearly all molecules $\theta_r/T \ll 1$ unless T is quite small, so that (5.36) can be replaced by

$$q_r = \int_0^\infty (2J + 1) \exp[-J(J + 1)\theta_r/T]\, dJ \tag{5.37}$$

By making the substitution $x = J(J + 1)$, and therefore $dx = (2J + 1)dJ$, this integral is easily shown to give

$$q_r = \frac{T}{\theta_r} = \frac{8\pi^2 IkT}{h^2}$$

However, for linear molecules which have a center of symmetry, for example N_2 or $HC{\equiv}CH$, J cannot acquire all possible integral values—rather it can have only all odd or all even values. This means that the sum in (5.36) and the integral in (5.37) will be half as large. In order to have an expression which applies to both symmetric and unsymmetric linear molecules, therefore, the above expression for the partition function is written

$$q_r = \frac{T}{\sigma\theta_r} = \frac{8\pi^2 IkT}{\sigma h^2} \tag{5.38}$$

where σ, called the symmetry number, is unity for unsymmetric and two for symmetric linear molecules. The symmetry number is, in fact, the number of equivalent or indistinguishable orientations in space which the molecule can acquire as a result of its tumbling motion. A symmetric molecule will have two equivalent orientations because it can be turned end-over-end without altering its appearance. For linear molecules at ordinary temperatures q_r is usually of the order of 10^1 to 10^2.

In some situations, those in which I is very small and/or T is very small, θ_r/T is large enough to render replacing (5.36) by (5.37) invalid or only marginally acceptable. At 300 K, for example, the error in q_r resulting from the use of (5.37) for some small molecules is: H_2, 9%, HD, 7%; D_2, 5%; HCl, 2%. The error is larger, of course, at lower temperatures. A good rule is to use (5.37), and therefore (5.38), only when θ_r/T is less than or equal to 0.01. If $\theta_r/T > 0.01$, however, then it is better to evaluate q_r by term-by-term summation according to (5.36). Alternatively, but only when $0.5 > \theta_r/T > 0.01$, the "Mulholland approximation,"

$$q_r = \frac{T}{\sigma\theta_r}\left[1 + \frac{1}{3}\left(\frac{\theta_r}{T}\right) + \frac{1}{15}\left(\frac{\theta_r}{T}\right)^2 + \frac{4}{315}\left(\frac{\theta_r}{T}\right)^3\right] \tag{5.39}$$

may be used to give a quick, accurate result. [For $\theta_r/T < 0.01$, (5.39) is valid but gives no better value than the simpler (5.38).]

EXAMPLE 5.6. (a) Compute q_r for CO_2 at 298.15, given that $I = 7.18 \times 10^{-46}$ kg m^2. (b) Repeat the calculation for HD at the same temperature. $I = 6.29 \times 10^{-48}$ kg m^2.

ANSWER. (a) From (5.38), since $\sigma = 2$,

$$q_r = \frac{8\pi^2(7.18 \times 10^{-46})(1.381 \times 10^{-23})(298.15)}{2(6.26 \times 10^{-34})^2} = 265.7$$

(b) For HD, $\theta_r/T = h^2/8\pi^2(6.29 \times 10^{-48})k(298) = 0.215$. We may therefore find q_r by either (5.39) or (5.36). Using (5.39) we have, since $\sigma = 1$,

$$q_r = \frac{1}{1 \times 0.215}\left[1 + \frac{1}{3}(0.215) + \frac{1}{15}(0.215)^2 + \frac{4}{315}(0.215)^3\right] = 5.00$$

Alternatively, using (5.36), we have

$$q_r = e^{-0} + 3e^{-2(0.215)} + 5e^{-6(0.215)} + 7e^{-12(0.215)} + \cdots = 5.00$$

[Had we used (5.38) we would have found $q_r = 4.65$, in error by 7%.]

Since (5.38) gives q_r for linear molecules in most cases to a very good approximation we shall use it in determining most rotational contributions to the thermodynamic functions. Proceeding as in Section 5.1 we use (4.34) to determine E_r:

$$E_r = NkT^2 \frac{d}{dT} \ln \frac{T}{\sigma\theta_r} = NkT$$

or

$$E_r \text{ (per mole)} = RT \tag{5.40}$$

Furthermore, the rotational contribution to the heat capacity will be

$$C_r \text{ (per mole)} = R \tag{5.41}$$

These results are entirely in accord with classical theory.

The rotational contribution to the entropy is obtained, similarly, using (5.40) in (4.39):

$$S_r = kN \ln \frac{T}{\sigma\theta_r} + \frac{NkT}{T} = kN \ln \frac{T}{\sigma\theta_r} + kN \tag{5.42}$$

or

$$S_r \text{ (per mole)} = R \ln \frac{T}{\sigma\theta_r} + R \tag{5.43}$$

which, at 298.15 K, becomes

$$S_{r298} \text{ (J/K mol)} = 55.69 - R \ln \sigma\theta_r \tag{5.44}$$

Alternatively, in terms of I instead of θ_r, we may write

$$S_r \text{ (J/K mol)} = R \ln \frac{IT}{\sigma} + 877.38 \tag{5.45}$$

which, at 298.15 K, becomes

$$S_{r298} \text{ (J/K mol)} = R \ln \frac{I}{\sigma} + 924.76 \tag{5.46}$$

In (5.45) and (5.46) I must be in kg m^2.

EXAMPLE 5.7. (a) For $^{14}N_2$ the molecular weight is 28.01 g/mol and the bond length is 109.5 pm. Find the translational and rotational entropies per mole at 298.15 K and 1.000 atm. (b) The standard third law entropy of $^{14}N_2$ at this temperature is 192.0 J/K mol corrected to ideal behavior. Compare this with the results in (a) and draw conclusions.

ANSWER. (a) Using (5.27) and (5.28) $I = [(28.01/2L)^2/28.01/L)] \times (109.5 \times 10^{-12})^2 = 1.394 \times 10^{-43}$ g m^2 or 1.394×10^{-46} kg m^2. From (5.16) $S_t^0 = \frac{3}{2}R \ln 28.01 + 108.745 = 150.3$ J/K mol and, from (5.46), $S_r = R \ln (1.394 \times 10^{-46}/2) + 924.76 = 41.1$ J/K mol.

(b) Since $S_t^0 + S_r = 191.4$, which is close to the third law value of 192.0, it appears that the translational and rotational contributions are the only appreciable ones.

5.4. THE ROTATIONAL PARTITION FUNCTION FOR NONLINEAR MOLECULES

The vast majority of molecules are nonlinear. The derivation of the expression for q_r will not be undertaken because it is too complicated—only the final result will be given. Nonlinear molecules have three nonzero principal moments of inertia, although two, or all three, of them may be equal. Regardless of this it is found that

$$q_r = \frac{\pi^{1/2}}{\sigma}\left(\frac{8\pi^2 I_A kT}{h^2}\right)^{1/2}\left(\frac{8\pi^2 I_B kT}{h^2}\right)^{1/2}\left(\frac{8\pi^2 I_C kT}{h^2}\right)^{1/2} \tag{5.47}$$

The symmetry number, σ, has the same meaning as used earlier. Representative values of it are: 2 for H_2O, 12 for C_6H_6, 12 for CH_4.

Equation (5.47) can be written in other ways, for example,

$$q_r = \frac{(\pi I_A I_B I_C)^{1/2}}{\sigma}\left(\frac{8\pi^2 kT}{h^2}\right)^{3/2} \tag{5.48}$$

and

$$q_r = \frac{\pi^{1/2}}{\sigma}\left(\frac{T^3}{\theta_A \theta_B \theta_C}\right)^{1/2} \tag{5.49}$$

where the characteristic rotational temperatures, θ_A, θ_B, and θ_C, are defined by

$$\theta_A = \frac{h^2}{8\pi^2 I_A k}, \qquad \theta_B = \frac{h^2}{8\pi^2 I_B k}, \qquad \theta_C = \frac{h^2}{8\pi^2 I_C k} \tag{5.50}$$

which are analogous to the definition of θ_r for linear molecules. The reader is reminded that it is the product of the I's (or θ's) which is needed, not the individual values.

Substituting (5.48) in (4.34) leads to

$$E_r = kNT^2 \frac{d}{dT} \ln \left[\frac{\pi(I_A I_B I_C)^{1/2}}{\sigma} \left(\frac{8\pi^2 kT}{h^2} \right)^{3/2} \right]$$

which reduces readily to

$$E_r = \tfrac{3}{2}kNT \tag{5.51}$$

or

$$E_r \text{ (per mole)} = \tfrac{3}{2}RT \tag{5.52}$$

and the rotational heat capacity is

$$C_r \text{ (per mole)} = \tfrac{3}{2}R \tag{5.53}$$

Once again we have results, (5.52) and (5.53), which are those predicted by classical theory.

Combining (4.39), (5.48), and (5.51) yields

$$S_r = \tfrac{1}{2}kN \ln \left[\pi I_A I_B I_C \left(\frac{8\pi^2 kT}{h^2} \right)^3 \right] - kN \ln \sigma + \tfrac{3}{2}kN$$

or

$$S_r \text{ (per mole)} = \frac{R}{2} \ln \left[\pi I_A I_B I_C \left(\frac{8\pi^2 kT}{h^2} \right)^3 \right] - R \ln \sigma + \tfrac{3}{2}R \tag{5.54}$$

which reduces to

$$S_r\,(\mathrm{J/K\ mol}) = R \ln \frac{(I_A I_B I_C)^{1/2}}{\sigma} + \tfrac{3}{2}R \ln T + 1320.84 \qquad (5.55)$$

when the I's are in kg m^2. In terms of characteristic rotational temperatures this is the same as

$$S_r\,(\text{per mole}) = \frac{R}{2} \ln \frac{\pi T^3}{\theta_A \theta_B \theta_C} - R \ln \sigma + \tfrac{3}{2}R \qquad (5.56)$$

At 298.15 K (5.55) reduces to

$$S_{r298}\,(\mathrm{J/K\ mol}) = R \ln \frac{(I_A I_B I_C)^{1/2}}{\sigma} + 1391.89 \qquad (5.57)$$

and (5.56) to

$$S_{r298}\,(\mathrm{J/K\ mol}) = 88.289 - R \ln [\sigma(\theta_A \theta_B \theta_C)^{1/2}] \qquad (5.58)$$

EXAMPLE 5.8. Find the rotational energy, heat capacity, and entropy of $NH_3(g)$ at 298.15 K, given that its characteristic rotational temperatures are 14.303, 14.303, and 9.080 K. The molecule is pyramidal.

ANSWER. From (5.52) $E_r = \tfrac{3}{2}R(298.15) = 3718$ J/mol; from (5.53) $C_r = \tfrac{3}{2}R = 12.47$ J/K mol. From (5.58), since $\sigma = 3$, $S_r = 88.289 - R \ln [3(14.303 \times 14.303 \times 9.080)^{1/2}] = 47.864$ J/K mol. *Note:* Since NH_3 is a symmetric top two of its I's are equal and therefore two of its θ_r's are equal.

5.5. THE VIBRATIONAL PARTITION FUNCTION

We must now consider the third kind of energy which can have an influence on the values of the thermodynamic functions. If it is recalled that the statistical approach leads, with only a few exceptions as already noted, to the same results for the translational and rotational energies and heat capacities of ideal gases as had been found by classical theory, one may begin to wonder whether the statistical approach is worth the effort. We shall see, however, that the statistical approach to the vibrational energy does indeed provide a significant improvement over the classical.

A diatomic molecule has only one vibrational mode—the periodic movement of the two atoms towards and away from each other, as if they were held together by a spring. A polyatomic molecule can have many vibrational modes. If n is the number of atoms in the molecule the number of vibrational modes is given by $3n - 5$ if the molecule is linear and by $3n - 6$ if it is nonlinear. Thus for CO_2 (linear) there are $3(3) - 5 = 4$ modes whereas for SO_2 (nonlinear) there are $3(3) - 6 = 3$ modes, and for C_6H_6 there are 30 modes. The four modes for CO_2 are as follows:

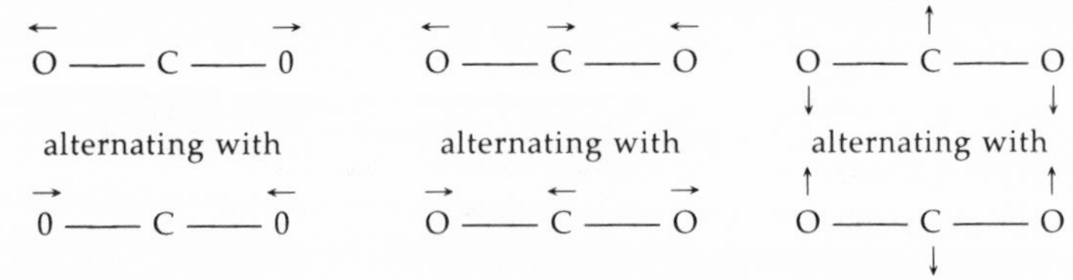

(1)	(2)	(3)
← →	← → ←	↑ (C)
O —— C —— 0	O —— C —— O	O —— C —— O
		↓ (O) ↓ (O)
alternating with	alternating with	alternating with
→ ←	→ ← →	↑ (O) ↑ (O)
0 —— C —— 0	O —— C —— O	O —— C —— O
		↓ (C)
(1) Symmetric stretch	(2) Asymmetric stretch	(3) Bending

The fourth mode, not shown, is the same as (3) except that the bending is in a plane perpendicular to the plane of the paper instead of in that plane. Each mode is characterized by a certain frequency, ν, as follows: $\nu_{(1)} = 3.939 \times 10^{13}$, $\nu_{(2)} = 7.000 \times 10^{13}$, and $\nu_{(3)} = \nu_{(4)} = 1.988 \times 10^{13}\ s^{-1}$. The two bending modes have identical frequencies, and one may say that the bending vibration is doubly degenerate. It is immaterial whether we consider (3) and (4) as two nondegenerate modes or as one doubly degenerate. We shall regard them as the former. Vibration frequency information is gained from spectroscopic measurements.

The vibrational motion of the atoms is thus quite complicated, and can involve not only stretching and bending of the molecule, but also twisting, breathing, group wagging, and other movements—all superimposed. Fortunately, however, all the modes are independent of each other as are the corresponding energies. The vibrational energy can therefore be regarded as the sum of the energies of each mode, that is,

$$\epsilon_v = \epsilon_{(1)} + \epsilon_{(2)} + \epsilon_{(3)} + \cdots \quad (5.59)$$

Since the energy *of each mode* is given by

$$\epsilon_{(i)} = (v + ½)\, h\nu_{(i)} \quad (5.60)$$

where v, the vibrational quantum number, is 0, 1, 2, . . . [compare (3.17)], we may write an expression for the partition function, q_v, for the ith mode as follows [compare (3.18)]:

$$q_{v(i)} = e^{-h\nu(i)/2kT} + e^{-3h\nu(i)/2kT} + e^{-5h\nu(i)/2kT} + \cdots \quad (5.61)$$

where the energy zero is the hypothetical vibrationless molecule. Alternatively, we may write

$$q_{v(i)} = e^{-0} + e^{-h\nu(i)/kT} + e^{-2h\nu(i)/kT} + \cdots \quad (5.62)$$

where the energy zero is the ground vibrational state.

Just as the separability of ϵ_t, ϵ_r, ϵ_v, etc. led to $q = q_t q_r q_v \ldots$ in (4.26), so the separability of the ϵ's of the various modes permits one to write

$$q_v = q_{v(1)} q_{v(2)} q_{v(3)} \cdots = \prod_i q_{v(i)} \quad (5.63)$$

and, of course,

$$\ln q_v = \sum_i \ln q_{v(i)} \tag{5.64}$$

It will be recognized that $\ln q_v$ is what enters into the computation of the contribution of vibration to the various thermodynamic functions, just as $\ln q_t$ and $\ln q_r$ did. However, rather than find $q_{v(i)}$ for each mode, multiply them all together to give q_v according to (5.63), and then take the logarithm, it is more convenient and enlightening to find $\ln q_{v(i)}$ for each mode and compute (where necessary) the contribution of that particular mode to the thermodynamic functions.

Returning to (5.61) we make the substitution

$$\theta_{v(i)} = h\nu_{(i)}/k \tag{5.65}$$

(compare Section 3.2), where $\theta_{v(i)}$ is the characteristic vibrational temperature of the given mode, and rewrite $q_{v(i)}$ as follows:

$$q_{v(i)} = e^{-\theta_{v(i)}/2T}[1 + e^{-\theta_{v(i)}/T} + e^{-2\theta_{v(i)}/T} + \cdots] \tag{5.66}$$

which is the same as

$$q_{v(i)} = e^{-\theta_{v(i)}/2T}[1 - e^{-\theta_{v(i)}/T}]^{-1} \tag{5.67}$$

This is the form of the vibrational partition function to be used when the energy zero is the vibrationless molecule. Had we performed the same operations on (5.62) we would have found that

$$q_{v(i)} = [1 - e^{-\theta_{v(i)}/T}]^{-1} \tag{5.68}$$

where the energy zero is the ground vibrational level (compare Problem 3.3). Calculated in this way $q_{v(i)}$ has an order of magnitude at ordinary temperatures of 10^0–10^1.

The overall vibrational partition function, q_v, is therefore given by

$$q_v = \prod_i e^{-\theta_{v(i)}/2T}[1 - e^{-\theta_{v(i)}/T}]^{-1} \tag{5.69}$$

or by

$$q_v = \prod_i [1 - e^{-\theta_{v(i)}/T}]^{-1} \tag{5.70}$$

depending on the choice of energy zero.

EXAMPLE 5.9. Use the data just given for CO_2 to compute, for each vibrational mode, the wave number, the characteristic vibration temperature, and the partition function at 298.15 K, based on energy zeros equal to the ground vibrational states. Compare the modes as to the distribution of population among the energy levels.

ANSWER. Since the wave number $\tilde{\nu} = \nu_{(i)}/c$ (where c is the velocity of light), use of (5.65) and (5.68) gives the results shown in Table 5.2. For the stretching modes, since q_v is close to unity, practically all the molecules are in the ground vibrational state at this temperature. For the bending modes there is a small but appreciable population of the first excited state.

Table 5.2. Vibration Parameters for CO_2

Mode	$\nu_{(i)}$ (s^{-1})	$\hat{\nu}_{(i)}$ (m^{-1})	$\theta_{v(i)}$ (K)	$q_{v(i)}$
(1)	3.939×10^{13}	1.314×10^5	1890	1.002
(2)	7.000×10^{13}	2.335×10^5	3360	1.000
(3)	1.988×10^{13}	6.63×10^4	954	1.042
(4)	1.988×10^{13}	6.63×10^4	954	1.042

We are now in a position to calculate the vibrational contributions to the state functions using (4.35), (4.40), (4.41), and (4.42). From (5.59) we can write for the vibrational energy of the whole sample

$$E_v = \sum_i E_{v(i)} \tag{5.71}$$

Applying (4.35) to each individual mode yields, after insertion of (5.67),

$$\begin{aligned} E_{v(i)} &= kNT^2 \frac{d}{dT} \ln [e^{-\theta_{v(i)}/2T}(1 - e^{-\theta_{v(i)}/T})^{-1}] \\ &= kNT^2 \left[\frac{\theta_{v(i)}}{2T^2} + \frac{e^{-\theta_{v(i)}/T}}{1 - e^{-\theta_{v(i)}/T}} (\theta_{v(i)}/T^2) \right] \end{aligned}$$

or

$$E_{v(i)} = ½kN\theta_{v(i)} + \frac{kN\theta_{v(i)}}{e^{\theta_{v(i)}/T} - 1}$$

so

$$E_{v(i)} \text{ (per mole)} = ½R\theta_{v(i)} + \frac{R\theta_{v(i)}}{e^{\theta_{v(i)}/T} - 1} \tag{5.72}$$

Use of (5.68) instead of (5.67) would have given

$$E_{v(i)} \text{ (per mole)} = \frac{R\theta_{v(i)}}{e^{\theta_{v(i)}/T} - 1} \tag{5.73}$$

Both (5.72) and (5.73) are equivalent to

$$E_{v(i)} - E_{0v(i)} = \frac{R\theta_{v(i)}}{e^{\theta_{v(i)}/T} - 1} \text{ per mole} \tag{5.74}$$

since $E_{0v(i)} = \frac{1}{2}R\theta_{v(i)}$ in (5.72) and zero in (5.73). It is recommended that the student reread the concluding portion of Section 4.2 at this point.

Finally, the overall vibrational energy is given, from (5.71), by

$$E_v - E_{0_v} = \sum_i (E_{v(i)} - E_{0v(i)}) = \sum_i \frac{R\theta_{v(i)}}{e^{\theta_{v(i)}/T} - 1} \text{ per mole} \quad (5.75)$$

We pause to note that as T becomes indefinitely large, $E_{v(i)} - E_{0v(i)}$ in (5.73) approaches RT, the classical value for an active vibrational mode. Since $E_{v(i)} - E_{0v(i)}$ at ordinary temperatures is rarely as large as RT it is clear that the classical result is only a limiting value. It was one of the triumphs of the present approach that the calculation of correct energy values became available.

The vibrational contribution *of each mode* to the molar heat capacity will be given, of course, by the derivative of $E_{v(i)}$ or of $E_{v(i)} - E_{0v(i)}$ with respect to T:

$$C_{v(i)} = \frac{dE_{v(i)}}{dT} = \frac{d}{dT}\frac{R\theta_{v(i)}}{e^{\theta_{v(i)}/T} - 1} = \frac{R(\theta_{v(i)}/T)^2 e^{\theta_{v(i)}/T}}{(e^{\theta_{v(i)}/T} - 1)^2} \quad (5.76)$$

EXAMPLE 5.10. (a) Find E_t, E_r, and E_v for one mole of $Cl_2(g)$ at 298.15 K assuming it to be ideal and taking the zero point energy as the energy zero. $\theta_v = 810$ K. (b) Find also the contributions of the same three forms of energy to the molar heat capacity, and compute C_P.

ANSWER. (a) By (5.10) and (5.40) $E_t = \frac{3}{2}R(298.15)$ and $E_r = R(298.15)$. By (5.73), since there is only one vibrational mode, $E_v = R(810)/(e^{810/298.15} - 1) = 57.32\ R$. Therefore $E_t + E_r + E_v = 802.7\ R = 6674$ J/mol.

(b) By (5.11), (5.41), and (5.76), $C_{V_t} + C_r + C_v = \frac{3}{2}R + R + (810/298.15)^2 e^{810/298.15} R/(e^{810/298.15} - 1)^2 = 25.43$ J/K mol. Therefore $C_P = 25.43 + R = 33.74$ J/K mol.

It can be easily shown that, for $\theta_{v(i)}/T > 6.5$, the contribution of the ith mode to C_v is less than 0.5 J/K mol, and therefore negligible for many purposes. This situation results when T is sufficiently small and/or $\theta_{v(i)}$ is sufficiently large. Since a large $\theta_{v(i)}$ is caused by a large value for ν_i which, in turn, is caused by a strong bond, we find that, at ordinary temperatures, only the modes associated with comparatively weak bonds contribute to C_v. Thus at room temperature the strong triple bond in N_2, for which $\theta_v = 3374$ K, relates to a mode which makes no contribution to C_v (see Example 5.7) whereas the relatively weak bond in I_2, for which $\theta_v = 307$ K, relates to a mode which contributes as much as $0.9R$ to C_v.

Proceeding further we can find the overall vibrational entropy, S_v, which is equal to $\Sigma_i S_{v(i)}$. With the help of (4.40) we write

$$S_v = \sum_i \left(kN \ln q_{v(i)} + \frac{E_{v(i)}}{T} \right) \tag{5.77}$$

If we use (5.68) for $q_{v(i)}$ and (5.73) for $E_{v(i)}$ we find

$$S_{v(i)} = \frac{kN(\theta_{v(i)}/T)}{e^{\theta_{v(i)}/T} - 1} - kN \ln (1 - e^{-\theta_{v(i)}/T}) \tag{5.78}$$

Had we used (5.67) for $q_{v(i)}$ and (5.72) for $E_{v(i)}$ we would have obtained an identical result so that the choice of energy zero does not affect the entropy. (A little thought will show why this is to be expected.) It follows, of course, from (5.78), that the overall vibrational entropy is given by

$$S_v = \sum_i \left[\frac{kN(\theta_{v(i)}/T)}{e^{\theta_{v(i)}/T} - 1} - kN \ln (1 - e^{-\theta_{v(i)}/T}) \right] \tag{5.79}$$

or

$$S_v \text{ (per mole)} = \sum_i \left[\frac{R(\theta_{v(i)}/T)}{e^{\theta_{v(i)}/T} - 1} - R \ln (1 - e^{-\theta_{v(i)}/T}) \right] \quad (5.80)$$

EXAMPLE 5.11. Calculate the standard entropy of $^{35}Cl_2(g)$ at 298.15 K, given that $M = 69.94$ g/mol, $\theta_r = 0.351$ K, and $\theta_v = 810$ K.

ANSWER. By (5.16) $S_t^0 = (\frac{3}{2})R \ln 69.94 + 108.745 = 161.72$ J/K mol; by (5.44) $S_r = 55.69 - R \ln (2 \times 0.351) = 58.63$ J/K mol. Since $\theta_v/298.15 = 2.717$, use of (5.80) leads to $S_v = 2.717\ R/(e^{2.717} - 1) - R \ln (1 - e^{-2.717}) = 2.17$ J/K mol. Therefore $S_{298}^0 = S_t^0 + S_r + S_v = 222.52$ J/K mol.

There remains the contribution of vibrational energy to the Helmholtz and Gibbs free energies, both of which are given by $-RT \ln q_v$ per mole according to (4.41) and (4.42). With energy zero assigned to the hypothetical vibrationless molecule, use of (5.69) for q_v gives

$$\begin{aligned} A_v \text{ (per mole)} &= G_v \text{ (per mole)} \\ &= \sum_i [\tfrac{1}{2}R\theta_{v(i)} + RT \ln (1 - e^{-\theta_{v(i)}/T})] \end{aligned} \quad (5.81)$$

or

$$\begin{aligned} A_v - A_{0_v} &= G_v - G_{0_v} \\ &= \sum_i RT \ln (1 - e^{-\theta_{v(i)}/T}) \text{ per mole} \end{aligned} \quad (5.82)$$

where

$$A_{0_v} = G_{0_v} = \sum_i \tfrac{1}{2}R\theta_{v(i)} \quad (5.83)$$

With energy zero equal to the ground vibrational level use of (5.70) for q_v gives the same result as (5.82) but

$$A_{0_v} = G_{0_v} = 0 \tag{5.84}$$

5.6. FURTHER COMMENTS ON ENERGY ZEROS

In the previous section, and in Section 3.2, reference has been made to the choice of different energy zeros for vibrational motion. For translational motion the ground-state energy is effectively zero even though the smallest value each translational quantum number, n, in (5.1) can acquire is unity and not zero—for when each n is unity ϵ is still no larger than 10^{-42} J in a vessel 1 dm^3 in volume, which is many orders of magnitude smaller than the average ϵ. For this reason, all the translational ϵ's, and the resulting E_t in (5.10), are not only energies above ground but absolute translational energies. Similarly, for rotational motion the ground-state energy is also zero because $J = 0$ in relations such as (5.33), and the resulting E_r in (5.40) and (5.52) are not only energies above ground but absolute rotational energies.

For vibrational motion, however, various alternatives present themselves, a full appreciation of which will have to await a study of chemical reactions. Figure 5.4 denotes schematically the variation of the potential energy of a diatomic molecule, E_v, as a function of the interatomic distance, r. As the atoms vibrate in, say, the $v = 2$ level, r increases from r_a to r_b and decreases back again to r_a, while the potential energy falls and rises, then falls and rises again. When $r = r_e$ the potential energy which the system had at the ends of the swing has converted to kinetic energy and is minimal,

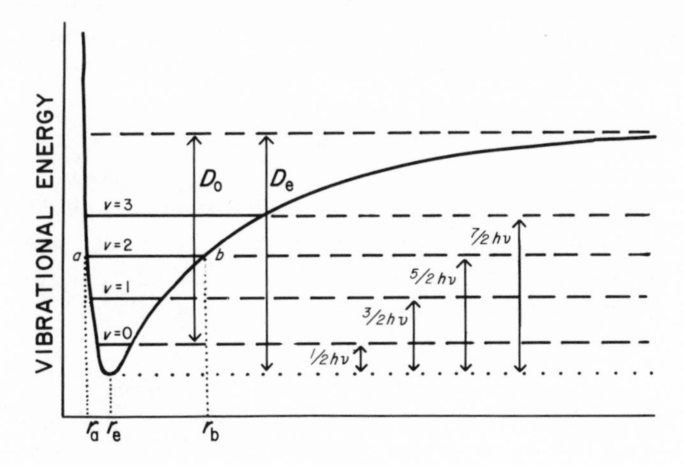

Figure 5.4. Potential energy of diatomic molecule as a function of bond length.

whereas when $r = r_a$ or r_b the potential energy is maximal and the kinetic energy zero. The energy in the ground state ($v = 0$) is $\frac{1}{2}h\nu$ above the energy of the hypothetical motionless system at the bottom of the potential well ($r = r_e$) but D_0 below that of the disassociated molecule—the energy approached asymptotically as $r \rightarrow \infty$. D_0 is the chemical dissociation energy of the bond of the diatomic molecule; D_e is called the spectroscopic dissociation energy.

When the energy of the hypothetical motionless system is assigned a value of zero the vibrational energy levels are given by $(v + \frac{1}{2})\, h\nu$, and $\epsilon_0 = \frac{1}{2}h\nu$, E_0 (per mole) $= \frac{1}{2}R\theta_v$. However, when the energy of the ground level is assigned a value of zero the vibrational energy levels are given by $vh\nu$, and $\epsilon_0 = 0$, $E_0 = 0$. But for both choices the energies above ground are the same, namely, $vh\nu$, with $E_v - E_0$ being given by (5.74).

A third choice for energy zero, the purpose of which will be shown in Chapter 7, is the system in the condition where the bond is completely dissociated. The vibrational energy levels are now given by $vh\nu - D_0$, a negative quantity, and $\epsilon_0 = -D_0$, E_0 (per mole) $= -LD_0$. But the energies above ground are still those given for the other choices for energy zero.

The use, therefore, of $E - E_0$, $H - H_0$, $A - A_0$, or $G - G_0$, whether referring only to the contributions of individual kinds of energy or to the overall values, is independent of the choice of energy zero provided E_0, H_0, A_0, and G_0 are the lowest accessible values in the various categories—the ones possessed by all the molecules at 0 K.

One last but important point should be made. Recalling the definitions of $H = E + PV$ (which equals $E + RT$ per mole for ideal gases) and of $A = E - TS$ and $G = H - TS$, it is clear that at 0 K, where all the molecules in the (hypothetical) gas state possess the lowest possible energies, $E = H = A = G$. Therefore, for ideal gases, $E_0^0 = H_0^0 = A_0^0 = G_0^0$. It must be understood, therefore, that for them $H^0 - E_0^0 = H^0 - H_0^0$ and $G^0 - E_0^0 = G_0 - H_0^0 = G_0 - G_0^0$, etc. These various ways of expressing the same quantity are commonly found, and can be a source of confusion if not understood.

PROBLEMS

5.1. (a) Calculate the molar translational entropy at 298.15 K and 1 atm of (i) Zn(g), $M = 65.37$ and (ii) HCl(g), $M = 36.46$ g/mol.

(b) The experimental (third law) entropies are 160.7 and 186.2 J/K mol, respectively. Draw conclusions.

***5.2.** (a) The Sackur–Tetrode equation may be written S (per mole) $= R(\ln V + \frac{3}{2} \ln T + \frac{3}{2} \ln M) + C$, where C is a constant. Evaluate C when this equation is used to find the molar entropy in SI units, but with M in the usual units of g/mol.

(b) Use your answer to (a) to find the molar entropy of Ar(g), $M = 39.948$ g/mol, at 1 atm and 298.15 K, and compare your result with the calculated value given in Example 5.2.

5.3. Locate the center of mass for the chloromethyl radical (CH_2Cl) assuming it to be planar with all bond angles 120°. Take the C–H and C–Cl bond lengths to be 110 and 170 pm, respectively, and the atomic weights to be H = 1.0, C = 12.0, and Cl = 35.0 g/mol.

5.4. (a) Use the data and result of Problem 5.3 to determine the orientation of the principal axes of the chloromethyl radical.

(b) Find the three principal moments of inertia and multiply them together.

(c) Confirm the choice of principal axes by showing that the products of inertia are all zero, according to (5.24), (5.25), and (5.26).

(d) Confirm that $I_A + I_B = I_C$ for this planar species.

***5.5.** Determine the moment of inertia of the (linear) OCS molecule without locating the center of mass, given that the atomic weights are C = 12.0, O = 16.0, S = 32.1 g/mol, and that the C–O and C–S bond lengths are 116 and 156 pm, respectively.

5.6. (a) For the $^{1}H^{16}O^{2}D$ molecule the following accurate data are available: atomic masses: $^{1}H = 1.0078$, $^{2}D = 2.0141$, $^{16}O = 15.9949$ amu; H–O–D bond angle = 104.53°; H–O and D–O bond lengths both equal to 95.71 pm. Locate the center of mass of the molecule.

(b) Evaluate $I_A I_B I_C$.

***5.7.** Find $I_A I_B I_C$ for the $^{12}C^{1}H^{19}F^{35}Cl^{79}Br$ molecule assuming that all the bond angles are tetrahedral (109.47°), that the bond lengths are C–H = 109, C–F = 135, C–Cl = 177, C–Br = 194 pm, and that the atomic weights are

1.00, 12.00, 19.00, 35.00, and 79.00 g/mol for H, C, F, Cl, and Br, respectively.

5.8. (a) For N_2O, a linear molecule, $\theta_r = 0.602$ K. Find its moment of inertia. ($M = 44.010$ g/mol.)
(b) Compute its translational and rotational entropies at 298.15 K and 1 atm.
(c) The literature value for the standard entropy at this temperature is 220.0 J/K mol. Draw conclusions.

5.9. The gases CO and N_2 have nearly the same molecular weight and moment of inertia. Compare their translational and rotational entropies at the same temperature and pressure.

5.10. (a) For H_2, for which $\theta_r = 87.5$ K and $M = 2.016$ g/mol, (5.38) gives an error in q_r of about 10% at 300 K as already stated, and should be replaced by (5.39). Similarly (5.40) to (5.46) will be in error. Write out the expression for q_r at 298.15 K, term by term, for the first seven terms, and find an expression for E_r using (4.34).
(b) Use the expression to evaluate E_r per mole at 298.15 K and compare with the classical value given by (5.40).
(c) Compute S_t^0 and S_r per mole at 298.15 K using (5.39) for q_r and (4.39) for S_r and compare their sum with the literature value, 130.59 J/K mol, for the entropy of H_2(g).
(d) Repeat the computation in (c) using (5.38) for q_r and comment on the errors introduced in E_r and S_r by using (5.38) for q_r.

***5.11.** (a) Find the standard translational and rotational entropies of HOD(g) per mole at 373.15 K using the data and results of Problem 5.6.
(b) Would you expect S_t and S_r for H_2O to be larger or smaller than the values just found? Why?
(c) Making the reasonable assumption that only the

translational and rotational energies contribute appreciably to the thermodynamic properties of water vapor under these conditions predict the (total) molar entropy and heat capacity at constant pressure of HOD(g) at this temperature and a pressure of 1 atm.

5.12. Without resorting to arguments involving quantum numbers explain the fact that as the symmetry number increases the entropy decreases, other things being equal. (Hint: Use the Boltzmann–Planck equation.)

***5.13.** The nitrous oxide molecule (N_2O) is linear. Find C_V per mole for the gas at 293.0 K given the following characteristic vibrational temperatures: 3200, 1840, 850(2) K. (The experimental value is 30.01 J/K mol.)

5.14. (a) In comparing (3.20) with (5.72), (3.23) with (5.73), and (3.24) with (5.76) it is evident that one result is three times the other. Why? Are the particles to which the second of each pair of equations refers distinguishable or indistinguishable?

(b) Why are the θ's used in Chapter 3 generally smaller than the θ_v's used in Chapter 5?

***5.15.** Find the ideal value for $H - H_0$ for Br_2(g) at 298.15 K if its characteristic vibrational temperature is 463 K.

***5.16.** Calculate $(H^0_{298} - H^0_0)/298.15$, $-(G^0_{298} - H^0_0)/298.15$, C^0_{P298}, and S^0_{298} for cyclopropane(g) from the following data: Atomic weights: H = 1.008, C = 12.011; bond lengths: C–H = 108.9, C–C = 151.0 pm; bond angles C–C–C = 60.0°, H–C–H = 115.1°; fundamental vibration frequencies (cm^{-1}): 739(2), 852, 866(2), 1028(2), 1078, 1133, 1188(3), 1442(2), 1454, 3025(2), 3038, 3082(2), and 3101.

Chapter Six

THERMODYNAMIC FUNCTIONS FOR IDEAL GASES—PART II

In the previous chapter the contributions of translational, rotational, and vibrational energies to the state functions, which are the major ones in most cases, were described. In the present chapter the remaining contributions are considered.

6.1. THE ELECTRONIC PARTITION FUNCTION

For most molecules there is no contribution of electronic energy to the thermodynamic properties at ordinary temperatures. The permitted values of the electronic energy are normally so far above ground that virtually only the ground level is occupied, except at high temperatures. (Nitric oxide and the monatomic halogens are exceptions to this statement.) Nevertheless, the electronic partition function, q_e, given by

$$q_e = g_0 e^{-\epsilon_0/kT} + g_1 e^{-\epsilon_1/kT} + g_2 e^{-\epsilon_2/kT} + \cdots \tag{6.1}$$

is not necessarily unity, even when ϵ_0 is taken as zero, because the electronic states may be degenerate. In some instances g_0 is not unity. This is especially true for molecules with an odd number of electrons, as for NO_2, NO, Na, and the monatomic halogens, but it is also true for O_2 with its two unpaired electrons. For O_2, for example, $g_0 = 3$. Taking the ground level as energy zero the first term in (6.1) is thus 3, even though all the remaining terms may be practically zero, and $q_e = 3$ at ordinary temperatures. Moreover, q_e is likely to be nearly independent of temperature (unless the latter is quite large). It follows that $d \ln q_e / dT \cong 0$, and the electronic contributions to E, H, C_V, and C_P are all zero. The entropy, however, depending as it does on $\ln q_e$ and not on its derivative, is affected. Under these circumstances $q_e = g_0$, so, by analogy with (4.39),

$$S_e \text{ (per mole)} = R \ln g_0 \qquad (6.2)$$

A similar contribution is made to the free energies:

$$A_e \text{ (per mole)} = G_e \text{ (per mole)} = -RT \ln g_0 \qquad (6.3)$$

It may be noticed that (6.2) is the same as would have resulted if W_{tot} in the Boltzmann–Planck equation had been replaced by g_0^L.

EXAMPLE 6.1. Find the standard molar entropy of $O_2(g)$ at 298.15 K given that $M = 32.00$ g/mol, $\theta_r = 2.07$ K, and $\theta_v = 2256$ K, and that the ground electronic level is triply degenerate.

ANSWER. By (5.16) $S_t^0 = \frac{3}{2}R \ln 32.00 + 108.745$; by (5.44) $S_r = 55.69 - R \ln (2 \times 2.07)$; by (5.80) $S_v \cong 0$ (since $\theta_v/298.15$ is so large); by (6.2) $S_e = R \ln 3$. The total of these contributions is $S_{298}^0 = 205.0$ J/K mol. *Note:* Omission of the electronic contribution would have given an error of nearly 5%.

Third law or calorimetric entropies may or may not include the ground electronic degeneracy contribution. The calculated (statistical) entropies are therefore considered more reliable, and are the ones used in tabulations of data.

An accurate treatment of the electronic contribution is often more complicated than indicated above, because electronic excitation can cause alterations in the rotational energies—a circumstance excluded at the outset in Section 4.3. The treatment of complications such as this is beyond the scope of this book.

6.2. RESIDUAL ENTROPY

With our consideration of translational, rotational, vibrational and electronic energies we have covered all the kinds of energy which are ordinarily of importance for calculations of the thermodynamic functions for "rigid" molecules. With the possible exception of the ground electronic degeneracy contribution, standard entropies calculated in this way are, or should be, the same as those obtained by the third law method, after correcting the latter to ideal behavior. Although the third law values are never greater than the calculated ones they are sometimes smaller, and the difference is more than can be accounted for by experimental error. Such a difference is called residual entropy.

The existence of residual entropy suggests that, even at 0 K, where none of the molecules has translational or rotational motion, and where they have only the zero point vibrational energy, there is still something contributing to degeneracy, so that $W_{tot} \neq 1$. This situation is found with CO, NO, N_2O, CH_3D, H_2O, long chain 1-olefins, and certain crystalline hydrates. In CO, for example, the standard third law entropy at 298.15 K amounts to 193.3 J/K mol whereas the statistical value is 198.0, giving a residual entropy of 4.7. The accepted explanation for the discrepancy is the following: At 0 K all the molecules in CO(s) should be oriented in the crystal lattice according to the strict pattern dictated by the unit cell. However, the C and O ends of the molecule are so similar—for example, they have such similar charge density distributions and dimensions—that the lattice will "accept" either a CO or an OC orientation without disruption. The result is that when CO(g) is cooled and solidified half of the molecules are "correctly" oriented while the others are oriented in the opposite direction, and the two "forms" are randomly distributed. In other words, there is a frozen-in disorder which is inevitable, and leads to $W = 2$ per molecule or 2^L per mole. By (2.7), therefore, $S = R \ln 2 = 5.7$ J/K mol, not greatly different from the 4.7 noted above. Since this explanation "overaccounts" for the observed result one may argue that the distribution of the two orientations is largely but not completely random. A similar explanation is given for the residual entropy found with N_2O, 4.9 J/K mol.

EXAMPLE 6.2. The residual entropy of NO(g) is 3.1 J/K mol, somewhat less than that for N_2O and CO. Account approximately for this value, supposing that NO is completely dimerized at low temperatures.

ANSWER. One mole NO $= L/2$ molecules of dimer, $(NO)_2$. If two equally probable orientations of this dimer are assumed, $W = 2^{L/2}$ so $S = k \ln 2^{L/2} = 2.88$ J/K mol at 0 K, which is close to the observed value.

Another kind of residual entropy is the presence of more than one isotopic form of an element. In the strict sense an element is not pure and therefore a compound of it is not pure if more than one isotope is present. Even at 0 K, therefore, the entropy of the crystalline form of such a compound is greater than zero by the entropy of mixing of the isotopic forms. Since, however, the proportion of the isotopes of every element remains virtually unchanged in any chemical reaction (with a few exceptions), the amount of such residual entropy is the same for reactants and products and therefore cancels in the determination of ΔS. For this reason such residual entropies are always ignored except in special cases involving isotope fractionation. It may be noted also that in most computations of moments of inertia ordinary chemical atomic weights (which are weighted averages of the isotopes) are used unless one is concerned with specific isotope species—even though moment of inertia is sensitive to atomic mass. The fact that bond lengths and bond angles are almost independent of isotopic content makes this practice acceptable.

6.3. THE NUCLEAR PARTITION FUNCTION

The treatment of nuclear energy resembles somewhat that of electronic energy in that (1) a large amount of energy is required to produce excited states and (2) these levels are often degenerate, even in the ground level. So much energy is required to excite nuclei to higher energy levels that only the ground levels are occupied except at exceedingly high temperatures.

The degeneracy of the ground level is determined by the nuclear spin quantum number i which, in turn, is determined by the number of protons (Z) and of neutrons (N) in the nucleus of the element being considered. With Z and N both even, $i = 0$ (as with ^{12}C and ^{16}O); with Z even and N odd, or with Z odd and N even, i is half-integral (for example, ½ for ^{1}H, 3/2 for ^{35}Cl, 5/2 for ^{17}O); with Z and N both odd, i is integral (for example, 1 for ^{2}D and ^{14}N, 3 for ^{10}B). The degeneracy is given by $2i + 1$, which applies to every nucleus in the species being considered. The nuclear degeneracy of the molecule as a whole, then, will be $\Pi_m (2i + 1)$, where m is the number of atoms in the species. Since the nuclear energy (above zero) is zero the nuclear partition function, q_n, will simply be the multiplicity in the ground state or

$$q_n = \prod_m (2i + 1) \tag{6.4}$$

and the nuclear entropy, S_n will be, by analogy with (6.4),

$$S_n \text{ (per mole)} = R \sum_m \ln (2i + 1) \qquad (6.5)$$

EXAMPLE 6.3. What is the molar nuclear entropy of $^1H^{12}C^{14}N$?

ANSWER. For H, C, and N, i = ½, 0, and 1, respectively. Therefore, by (6.5),

$$\begin{aligned} S_n &= R\{\ln [2(1/2) + 1] + \ln [2(0) + 1] + \ln [2(1) + 1]\} \\ &= 14.90 \text{ J/K mol} \end{aligned}$$

The contribution of the ground level nuclear degeneracy is not usually detected in third law measurements of entropy, and is normally ignored. Furthermore, it is generally independent of the manner in which the element is chemically bound. It is therefore the same for the elements of reactants as it is for those of the products, so that it has no effect on the changes in the thermodynamic functions for a given chemical reaction. For this reason it is excluded from tabulated values of state functions. One may wonder, then, why it is even necessary to discuss the subject. The answer is that for H_2 and D_2 a consideration of nuclear degeneracy is necessary to account for their unique behavior. This will be examined in the following section.

6.4. ORTHO AND PARA H_2 AND D_2

If the third law standard entropy for H_2 at 298.15 K, 124.0 J/K mol, found by measuring its heat capacity in the usual manner down to near 15 K and extrapolating to 0 K, is compared with that found by statistical calculations (see Problem

(5.10), 130.6 J/K mol, the difference between the two values is evident, and it is apparent that this residual entropy cannot be accounted for by the simple arguments of Section 6.2. The behavior of H_2 is actually considerably more complicated—more so than might be inferred from the wording of Problem 5.10. When, as with H_2 and D_2, two identical atoms whose nuclei have nonzero spin quantum numbers are bound in a diatomic molecule the existence of nuclear spin isomerism arises from symmetry considerations. For nuclei of odd mass number, such as H_2, quantum mechanics tells us that the combined rotational and spin wave functions must be antisymmetric with respect to the nuclei; for nuclei of even mass number, such as D_2, the combined wave functions must be symmetric. For H_2 this turns out to mean that when the rotational quantum numbers, J, are odd there are three nuclear quantum states and when the J's are even there is only one quantum state. These two kinds of H_2, both present in the ordinary gas, are described as ortho (J odd) and para (J even) isomers, respectively, so called because, in an approximate description, one may say that the nuclear spins of the two H atoms are parallel in the ortho but antiparallel in the para isomer.

Because of the connection between the nuclear and rotational symmetries it is usual to combine the nuclear and rotational degeneracies, and so write q_r for H_2 as follows:

$$q_r = \underbrace{3 \sum_{J \text{ odd}} (2J + 1)e^{-J(J+1)\theta_r/T}}_{\text{(ortho)}} + \underbrace{1 \sum_{J \text{ even}} (2J + 1)e^{-J(J+1)\theta_r/T}}_{\text{(para)}} \tag{6.6}$$

At high enough temperatures each of these sums is virtually equal to $T/2\theta_r$ so $q_r = 4(T/2\theta_r)$, or four times the value given by (5.38) used earlier. The four is just that factor expected from

(6.4) when $i = \frac{1}{2}$ and $m = 2$. It will also be evident that, at sufficiently high temperatures (at and above 300 K), there will be three times as many ortho terms in q_r as para, so that the mole fraction of ortho, X_{ortho}, should be 0.75 (see Section 1.4). At sufficiently low temperatures *all* the terms in the first sum of (6.6) should vanish leaving only the second sum. In other words, H_2 at 0 K should consist entirely of the one nuclear isomer of the form with $J = 0$, or $X_{\text{para}} = 1$ and $X_{\text{ortho}} = 0$. Thus the ratio of ortho to para decreases from a maximum of 3 to a minimum of 0 as T is reduced indefinitely.

It happens, however, that the shift in the ortho $\rightleftharpoons$ para equilibrium to the para side with cooling does not actually occur in the absence of a catalyst, so that the heat capacity measurements made on ordinary ("normal") hydrogen (ortho/para = 3) as it is cooled are measurements on the 3:1 mixture at *all* temperatures, a mixture which, of course, is metastable at the lower temperatures. When, therefore, the third law entropy of H_2 is measured in the usual manner down to about 12 K with extrapolation to 0 K it pertains to the 3:1 mixture as if it were a pure, nonisomeric substance. At 12 K all the para molecules are in the $J = 0$ state and the ortho in the $J = 1$ state. Third law measurements with extrapolation from 12 K will not therefore include any contribution from nuclear isomerism, from nuclear degeneracy, or from rotational degeneracy $(2J + 1 = 3)$ of the $J = 1$ ortho form. Obtained in this way the third law entropy is 124.0 J/K mol. To convert this into a value suitable for tabulation for use in conjunction with entropy data of other substances we must add the entropy due to the rotational degeneracy of the ortho form. Since there is only 0.75 of it the latter contribution will be $0.75\ R \ln 3 = 6.851$ J/K mol.

The so-called statistical entropy, however, is taken to include also the contributions of nuclear isomerism and nuclear degeneracy. These are the same as the entropy increase resulting when (1) 0.75 mole of the ortho form is formed by mixing 0.75/3 mole of each of the three ortho forms and (2) the resulting 0.75 mole of ortho is mixed with 0.25 mole of para. ΔS in (1) will be $0.75R \ln 3 = 6.851$ J/K mol, the same as the $J = 1$ rotational contribution. To find ΔS in (2) we recall that, from classical thermodynamics (or even from statistical thermodynamics), the entropy change on mixing to give an ideal solution is

$$\Delta S^M \text{ (per mole)} = -R \sum_i X_i \ln X_i \qquad (6.7)$$

where the sum includes all the components which are mixed. ΔS for (2) will hence be given by $-R(0.25 \ln 0.25 + 0.75 \ln 0.75) = 4.675$ J/K mol. The sum of ΔS in (1) and (2) is 11.526 J/K mol, the amount by which the tabulated entropy must be increased to give the statistical entropy (including nuclear spin contributions).

The statistical entropy of H_2 at 298.15 K and 1 atm from spectroscopic data and including nuclear spin contributions is calculated to be 142.12 J/K mol. Working backwards from this the tabulated entropy will be $142.12 - 11.526 = 130.59$ J/K mol, and the third law entropy, measured as described above, is expected to be $130.59 - 6.85 = 123.74$ J/K mol. The experimental value, subject to some error, is actually 124.01 J/K mol.

Ortho and para isomers are expected to occur with other homodiatomic molecules such as $^{14}N_2$, but their presence is not felt as it is with H_2. This is because in third law measurements of entropy their rotation ceases before the lower limit of experimental temperatures (at which C_P is measured) is reached. There is thus no residual rotational entropy.

EXAMPLE 6.4. Given that for H_2(g) $M = 2.0156$ g/mol, $\theta_r = 87.5$ K and $\theta_v = 6215$ K, find the statistical entropy at 80.0 K and 1 atm.

ANSWER. There will be no vibrational contribution since θ_v/T is so large, and no electronic contribution. By (5.15) $S_t^0 = 90.141$ J/K mol. By (6.6), since $\theta_r/T = 1.094$,

$$\begin{aligned} q_r &= 3(3e^{-2\theta_r/T} + 7e^{-12\theta_r/T} + \cdots) \\ &\quad + 1(1 + 5e^{-6\theta_r/T} + 9e^{-20\theta_r/T} + \cdots) \\ &= 2.016 \end{aligned}$$

Furthermore,

$$\begin{aligned} d \ln q_r/dT &= (1/q_r)(dq/dT) \\ &= (\theta_r/q_rT^2)\,[3(6e^{-2\theta_r/T} + 84e^{-12\theta_r/T} + \cdots) \\ &\quad + 30e^{-6\theta_r/T} + 180e^{-20\theta_r/T} + \cdots] \end{aligned}$$

$= 0.01398$ K^{-1}. Therefore, by (4.34) $E_r = 743.9$ J/mol and by (4.39) $S_r = 15.13$ J/K mol, and the statistical entropy is $90.14 + 15.13 = 105.27$ J/K mol. *Note:* The temperature is too low to use (5.38) even if the entropy excluding nuclear spin degeneracy were required.

6.5. INTERNAL ROTATION

We have thus far considered only molecules which, except for vibrational motion about equilibrium positions, are rigid. Many molecules do not conform to this requirement but undergo internal rotation about single bonds. The methyl group of toluene, for example, is nearly completely free to rotate about the bond which holds it to the aromatic ring. Similarly, rotation about the C–C bond in ethane is possible, but the rotation is more restricted than in toluene because of the proximity of the two methyl groups.

The presence of a freely rotating group reduces the number of vibrational modes by one so that the contribution to the thermodynamic functions of the missing vibrational mode is replaced by that of the free rotation, which has its own partition function, q_{ir}. When the rotating groups and the remainder of the molecule both have at least three-fold symmetry about the axis of internal rotation, the partition function for the free rotation is determined by the temperature and by the reduced moment of inertia, I_{ir}, defined by

$$I_{ir} = \frac{I_1 I_2}{I_1 + I_2} \tag{6.8}$$

where I_1 and I_2 are the moments of inertia of the two portions of the molecule (the two coaxial tops which are turning relative to each other) about the axis of internal rotation. For each such rotation, q_{ir} is then given by

$$q_{ir} = \frac{(8\pi^3 I_{ir} kT)^{1/2}}{\sigma_{ir} h} \tag{6.9}$$

where σ_{ir} is the number of times per complete rotation about the bond that the molecule returns to an equivalent position. It is analogous to, but different from σ, the symmetry number used for the overall rotation of the (rigid) molecules, used in (5.38) or (5.47). Thus σ_{ir} for the methyl group in Cl_3CCH_3 is 3.

EXAMPLE 6.5. Calculate q_{ir} for $Cd(CH_3)_2(g)$ at 298.15 K in which the two methyl groups are diametrically opposite each other and rotate freely about the C–C bond. The bond angles about the C atoms are all tetrahedral (109.5°), the C–H bond length is 109 pm, and the atomic weight of H is 1.008 g/mol.

ANSWER. Since the Cd and the two C atoms are all on the axis of internal rotation, the rotations of the two methyls can be considered to be a single rotation. Equation (6.8) is applicable because both portions of the molecule have three-fold symmetry about a common axis. Only the H's contribute to I_1, I_2, and I_{ir}. The perpendicular distance of each H atom from the axis of internal rotation is $109 \times \sin(180.0 - 109.5)$ pm. Each will therefore contribute $(1.008/L)[109 \times \sin(180.0 - 109.5)]^2 = 1.77 \times 10^{-20}$ g pm^2

to I_1 and I_2. Therefore $I_1 = I_2 = 3 \times 1.77 \times 10^{-20}$ g pm^2, according to (5.20), and I_{ir} from (6.8) $= 2.65 \times 10^{-20}$ g pm^2 $= 2.65 \times 10^{-47}$ kg m^2. Since $\sigma_{ir} = 3$,

$$q_{ir} = \frac{(8\pi^3 \times 2.65 \times 10^{-47} \times 1.38 \times 10^{-23} \times 298.15)^{1/2}}{3 \times 6.63 \times 10^{-34}} = 2.61$$

The energy due to each free rotation is obtained from $E_{ir} = RT^2(d \ln q_{ir}/dT)$ [see Eq. (4.34)] with the help of (6.9) to give

$$E_{ir} \text{ (per mole)} = \tfrac{1}{2}RT \tag{6.10}$$

It follows that the contribution to the heat capacity per rotating group is

$$C_{ir} \text{ (per mole)} = \tfrac{1}{2}R \tag{6.11}$$

and to the entropy per rotating group [see (4.39)] is

$$S_{ir} \text{ (per mole)} = R \ln q_{ir} + \tfrac{1}{2}R \tag{6.12}$$

The contribution to the Gibbs free energy is, by analogy with (4.42),

$$G_{ir} \text{ (per mole)} = -RT \ln q_{ir} \tag{6.13}$$

EXAMPLE 6.6. The sum of the translational, rotational (tumbling), and vibrational contributions to the standard entropy of $Cd(CH_3)_2(g)$ at 298.15 K is 290.5 J/K mol. Compute the contribution of free internal rotation of the methyl groups to the entropy, add it to the value just given, and compare the total with the third law standard entropy of 302.9 J/K mol $\pm$ 0.8.

ANSWER. Using (6.12) and the result of Example 6.5 we find $S_{ir} = R \ln 2.61 + \tfrac{1}{2}R = 12.1$, so $S^0_{298} = 290.5 + 12.1 = 302.6$ J/K mol, which agrees with the observed result well within experimental error.

The thoughtful reader may raise questions such as the following: Since the validity of (6.8) appears to be limited to internal rotations in which the two counterrotating portions of the molecule both have at least threefold symmetry, how does one treat some of the most common internal rotations such as those of the methyl groups in the various methylated benzenes where only the methyl group meets this requirement? What is the value of σ_{ir} for molecules like toluene, $C_6H_5CH_3$, if there is rotation about the C–CH_3 bond? If the CH_3 is held stationary while the C_6H_5 is turned does not one obtain $\sigma_{ir} = 2$, whereas if the C_6H_5 is held stationary while the CH_3 is turned down does not one obtain $\sigma_{ir} = 3$?

Somewhat superficial answers to these questions, but ones adequate for our purpose, can be given as follows: Even when the framework to which the rotating methyl group is attached does not have at least threefold symmetry, as with toluene, (6.8) may still be used as a good approximation as long as the moment of inertia of the internally rotating group is small compared with the principal moments of inertia. This is true for the methylated benzenes and for many other molecules. As for the values of σ and σ_{ir} for molecules like toluene it must be pointed out that it is the product of these two quantities, known as the complete symmetry number, $\sigma_{csn} = \sigma \cdot \sigma_{ir}$, which is of importance, not the individual values. It is clear that σ_{csn} is the total number of ways (equivalent positions) in which the molecule has the same spatial appearance, the "ways" resulting from *both* rotation of the molecule as a whole *and* the rotation of groups about bonds. Thus σ_{csn} is 6 for toluene. This can be seen by imagining the plane of the phenyl to be horizontal and allowing the methyl to rotate 360° (which gives three equivalent positions) and then turning the molecule over and again rotating the methyl (which gives another three, or a total of six ways). Having found σ_{csn} it can be used as shown below to replace both σ and σ_{ir}, thereby eliminating the difficulty or ambiguity that may accompany the assignment of the individual values.

Before showing how σ_{csn} is used, and as an aid to finding it, we now outline two approaches, using toluene once again for purposes of illustration. In the first method the internally rotating groups are locked in such a position as to give the molecule the maximum symmetry. For toluene this would mean one of the methyl hydrogens vertically above and the other two below the plane of the phenyl, the latter being horizontal. In this locked position $\sigma = 1$. By unlocking the methyl rotation and allowing it to rotate about the C–C bond there will be six equivalent positions, one every 60°, it being permitted to turn the molecule over. This gives $\sigma_{ir} = 6$, so $\sigma_{csn} = 1 \times 6 = 6$. In the second method each rotating methyl is replaced temporarily by a single atom, say a halogen (X) for the purpose of finding σ. For toluene this gives $\sigma = 2$. Restoring the methyls there will be three equivalent positions for each rotating methyl and 3^n for n rotating methyls, so $\sigma_{ir} = 3^n$. For toluene this procedure gives $\sigma_{ir} = 3^1 = 3$, so $\sigma_{csn} = 2 \times 3 = 6$. We have thus reached the same value for σ_{csn} by different routes. (In applying the second method to ethane *both* methyls must be replaced by X to find σ.) Typical values of σ_{csn} are: ethane(18), phenol(2), *p*-xylene(36).

Such values can now be used as follows to determine the *combined* tumbling rotation and internal rotation partition functions, $q_r(q_{ir})_1(q_{ir})_2 \ldots$, where the subscript numerals refer to the various internally rotating groups. From (5.49) for the molecule as a whole, and from (6.9) for each rotating group, we have

$$q_r(q_{ir})_1(q_{ir})_2 \ldots = \frac{\pi^{1/2}}{\sigma}\left(\frac{T^3}{\theta_A\theta_B\theta_C}\right)^{1/2}\left[\frac{(8\pi^3 I_{ir}kT)^{1/2}}{(\sigma_{ir})_1 h}\right]_1 \left[\frac{(8\pi^3 I_{ir}kT)^{1/2}}{(\sigma_{ir})_2 h}\right]_2 \ldots$$

$$= \pi^{1/2}\left(\frac{T^3}{\theta_A\theta_B\theta_C}\right)^{1/2}\left[\frac{(8\pi^3 I_{ir}kT)^{1/2}}{h}\right]_1$$

$$\left[\frac{(8\pi^3 I_{ir}kT)^{1/2}}{h}\right]_2 \cdots \frac{1}{\sigma(\sigma_{ir})_1(\sigma_{ir})_2 \cdots}$$

This can be written

$$q_r(q_{ir})_1(q_{ir})_2 \ldots = q_r'(q_{ir}')_1\,(q_{ir}')_2 \ldots (1/\sigma_{csn})$$

where the quantities with primes are, respectively, the previously defined partition functions *with all the symmetry numbers omitted.* Then, for example, the total entropy contribution of overall and internal rotation can be found from

$$\begin{aligned} &S_r + S_{ir}\ \text{(per mole)} \\ &\quad = R\ln q_r'(q_{ir}')_1(q_{ir}')_2 \ldots - R\ln\sigma_{csn} + \tfrac{3}{2}R + \tfrac{1}{2}mR \qquad (6.14) \end{aligned}$$

for a molecule with m freely rotating groups. Thus by finding σ_{csn} we have eliminated any difficulty in factoring it into σ and σ_{ir} by eliminating the need to do so.

Most internal rotations, however, are not free, but are restricted: there are potential energy barriers that must be surmounted by the rotating groups before they can complete a whole revolution. In ethane, for example, there are three such barriers. It is beyond the scope of this book to go into greater detail, except to say that the height of these barriers can be obtained from spectroscopy and, knowing them and q_{ir} calculated as above for free rotation, we can compute the contribution of the restricted rotation to the thermodynamic properties. Alternatively, the total of the calculated translational, rotational, and vibrational contributions can be subtracted from the measured value of a property and the difference, attributed to restricted rotation, used to compute the energy barrier. The presence of a barrier can increase or decrease the heat capacity and the enthalpy over that calculated for free rotation; the free energy is always increased (algebraically) over that for free rotation and the entropy is always decreased.

6.6. THE FREE ENERGY FUNCTION. TABULATION OF STATISTICAL THERMODYNAMIC DATA

Gibbs free energy data for elements and compounds, the most important kind of information for the practical applications of chemical thermodynamics, are usually tabulated as the (standard) free energy functions, $-(G_T^0 - H_0^0)/T$ or $-(G_T^0 - H_{298}^0)/T$, depending on the choice of reference state. When referring to gases the symbols with superscript zeros mean the values for the *ideal* gas state at a pressure of 1 atm; when referring to liquids and solids they mean those for the liquid or solid at a pressure of 1 atm. The subscripts give the temperature in Kelvins. The free energy function is chosen because it does not change markedly with temperature and so, by tabulating its values, for example at intervals of 500 K, it is possible to find it accurately at intermediate temperatures by linear interpolation. (In this it has the distinct advantage over free energy of formation which, although commonly tabulated for 298 K, is sensitive to temperature). Of the two functions just quoted, $-(G_T^0 - H_0^0)/T$ is the one most readily computed by statistical methods because energies are based on the values at 0 K. Another useful quantity is the enthalpy function $(H_T^0 - H_0^0)/T$, which, when T = 298.15 K, permits interconversion of the two free energy functions through the relation

$$-\frac{G_T^0 - H_{298}^0}{T} = \frac{H_{298}^0 - H_0^0}{T} - \frac{G_T^0 - H_0^0}{T} \tag{6.15}$$

As for the free energy function, $(H_T^0 - H_0^0)/T$ is less sensitive to temperature than is $H_T^0 - H_0^0$. The student is advised to reread the remarks at the end of Chapter 5 regarding the identity of G_0^0, H_0^0, and E_0^0 for gases.

We shall consider first the evaluation of $(H_T^0 - H_0^0)/T$. Since, of course,

$$\frac{H_T^0 - H_0^0}{T} = \left(\frac{H_T^0 - H_0^0}{T}\right)_t + \left(\frac{H_T^0 - H_0^0}{T}\right)_r + \left(\frac{H_T^0 - H_0^0}{T}\right)_v + \cdots$$

it is convenient to find the terms on the right and add them together. The first term is found from (5.12), the second from (5.40) or (5.52), and the third from (5.75). Notice that RT, which converts E to H, having been included in (5.12), is *not* included elsewhere. In other words, $H_{T_r}^0 = E_{T_r}^0$ and $H_{T_v}^0 = E_{T_v}^0$, etc.

EXAMPLE 6.7. Find $(H_T^0 - H_0^0)/T$ per mole at 500 K for $CO_2(g)$ by statistical methods, given that θ_v = 1890, 3360, 954, and 954 K (compare Example 5.9).

ANSWER. By (5.12) $[(H_{500}^0 - H_0^0)/500]_t = (5/2)R$ and, by (5.40), $[(H_{500}^0 - H_0^0)/500]_r = [(E_{500}^0 - E_0^0)/500]_r = R$. $[(H_{500}^0 - H_0^0)/500]_v = [(E_{500}^0 - E_0^0)/500]_v = (0.0883 + 0.0081 + 0.3324 + 0.3324)R = 0.7612R$, found by summing $R(\theta_{v(i)}/500)[e^{\theta_{v(i)}/500} - 1]$ according to (5.75) over every frequency. Therefore $(H_{500}^0 - H_0^0)/500 = (5/2)R + R + 0.7612R = 4.2612R = 35.43$ J/K mol.

The evaluation of $-(G_T^0 - H_0^0)/T$ follows a similar pattern. Since

$$-\frac{G_T^0 - H_0^0}{T} = -\frac{G_T^0 - E_0^0}{T}$$

$$= -\left(\frac{G_T^0 - E_0^0}{T}\right)_t - \left(\frac{G_T^0 - E_0^0}{T}\right)_r - \left(\frac{G_T^0 - E_0^0}{T}\right)_v - \cdots$$

we again find the individual contributions and add them together. Although, according to (4.42), the terms on the right are given by $R \ln (q_t^0/L)$, $R \ln q_r$, $R \ln q_v$, etc., it is more convenient to express them as follows: Since $G_T^0 = H_T^0 - TS_T^0$, or $-G_T^0/T = S_T^0 - (H_T^0/T)$,

$$-\left(\frac{G_T^0 - E_0^0}{T}\right)_t = S_t^0 - \left(\frac{H_T^0 - E_0^0}{T}\right)_t$$
$$= \tfrac{3}{2}R \ln M + \tfrac{5}{2}R \ln T - 9.685 - \tfrac{5}{2}R \text{ per mole}$$

with the help of (5.15) and (5.12), or

$$-\left(\frac{G_T^0 - E_0^0}{T}\right)_t (\text{J/K mol})$$
$$= \tfrac{3}{2}R \ln M + \tfrac{5}{2}R \ln T - 30.471 \quad (6.16)$$

with M in g/mol. Similarly, for *linear* molecules,

$$-\left(\frac{G_T^0 - E_0^0}{T}\right)_r (\text{per mole}) = S_r - \left(\frac{H_T^0 - E_0^0}{T}\right)_r$$
$$= R \ln \frac{IT}{\sigma} + 877.38 - R$$

or

$$-\left(\frac{G^0 - E^0}{T}\right)_r (\text{J/K mol}) = R \ln \frac{IT}{\sigma} + 869.07 \quad (6.17)$$

with the help of (5.45) and (5.40), I being in kg m^2. Alternatively,

$$-\left(\frac{G_T^0 - E_0^0}{T}\right)_r = R \ln \frac{T}{\sigma\theta_r} \quad (6.18)$$

by using (5.43) instead of (5.45). Note again that RT, which converts energy to enthalpy, having already been included in (6.16), is omitted in the remaining contributions such as (6.17) and (6.18). For *nonlinear* molecules (5.55) instead of (5.45), and (5.52) instead of (5.40) must be used, leading to

$$-\left(\frac{G_T^0 - E_0^0}{T}\right)_r = \tfrac{3}{2}R \ln T + R \ln \frac{(I_A I_B I_C)^{1/2}}{\sigma} + 1308.37 \text{ J/K mol} \qquad (6.19)$$

Similarly

$$-\left(\frac{G_T^0 - E_0^0}{T}\right)_v = \sum_i S_{v(i)} - \sum_i \left(\frac{H_T^0 - E_0^0}{T}\right)_v$$

Making the substitutions indicated in (5.75) and (5.80) results in simplification to

$$-\left(\frac{G_T^0 - E_0^0}{T}\right)_v = -R \sum_i \ln\left(1 - e^{\theta_{v(i)}/T}\right) \qquad (6.20)$$

The free energy functions for other forms of energy are handled in an analogous manner. The total free energy function is found by adding (6.16), (6.18), or (6.19), (6.20) and any others required.

EXAMPLE 6.8. Compute the standard free energy function per mole for CO_2(g) at 500 K, $-(G_{500}^0 - H_0^0)/500$, given the information in Example 6.7 and that $M = 44.00$ g/mol, $I = 7.18 \times 10^{-46}$ kg m^2.

(b) Combine this result with that of Example 6.7 to give S_{500}^0 per mole.

ANSWER. (a) Using (6.16) gives a translational contribution of 145.90 J/K mol. Using (6.17), with $\sigma = 2$ gives a rotational contribution of 50.71 J/K mol. The values of $-\ln(1 - e^{-\theta v(i)/T})$ for the four vibrational modes are, respectively, 0.0231, 0.0012, 0.1606, and 0.1606, totaling 0.3455 so, by (6.20), the vibrational contribution is 2.873 J/K mol. Adding the three contributions gives $-(G^0_{500} - H^0_0)/500 = 199.48$ J/K mol.

(b) $-(G^0_{500} - H^0_0)/500 + (H^0_{500} - H^0_0)/500 = 199.48 + 35.43 = 234.91$ J/K mol.

In order to find ΔG^0_T for a chemical change from values of the free energy function for each species occurring in the reaction we note that

$$\Delta G^0_T = \Delta H^0_0 - T\Delta\left(-\frac{G^0_T - H^0_0}{T}\right) \tag{6.21}$$

where Δ refers to the increment in the given property in the reaction. Clearly, a value for ΔH^0_0 is needed—the enthalpy change at 0 K for the given reaction. For reactions involving simple species, for example $I_2(g) \rightarrow 2I(g)$, this quantity is sometimes obtained from spectroscopy, but it is usually found by combining calorimetric values of ΔH^0_{298} with the enthalpy functions of the reactants and products according to

$$\Delta H^0_0 = \Delta H^0_{298} - (298.15)\Delta\left(\frac{H^0_{298} - H^0_0}{298.15}\right) \tag{6.22}$$

Alternatively, since

$$\Delta G^0_T = \Delta H^0_{298} - T\Delta\left(-\frac{G^0_T - H^0_{298}}{T}\right) \tag{6.23}$$

one can use ΔH^0_{298}, often determined directly from thermochemical experiments.

Finally, since $\Delta G^0_T = \Delta H^0_T - T\Delta S^0_T$,

$$\Delta S^0_{298} = \Delta\left(-\frac{G^0_{298} - H^0_{298}}{298}\right) \tag{6.24}$$

Table 6.1. Typical Free Energy Functions (Referred to 0 K)

	$-(G_T^0 - H_0^0)/T$, (J/K mol)			$H_{298}^0 - H_0^0$	ΔHf_0^0
	298.15 K	500 K	1000 K	(kJ/mol)	(kJ/mol)
$H_2(g)$	102.17	116.94	136.98	8.468	0
$I_2(g)$	226.69	244.60	269.45	10.117	65.10
$HI(g)$	177.40	192.42	212.97	8.657	28.0

so, if the free energy functions referred to 298 K are available, entropy changes can be found for that temperature.

In Tables 6.1 and 6.2 are presented a few data illustrating two ways in which they will be found in reference works of thermodynamic properties. The symbol ΔHf represents, of course, the standard enthalpy of formation from the elements in their conventional standard states at the temperature indicated by the subscript. Suppose we wish to find ΔG_{298}^0 and ΔG_{500}^0 for $H_2(g) + I_2(g) \rightarrow 2HI(g)$. Following the procedure of (6.21) and using the data of Table 6.1 gives

$$\Delta G_{298}^0 = 2(28.0) - 0 - 65.10$$
$$- \frac{298.15}{1000}[2(177.40) - 102.17 - 226.69] = -16.8 \text{ kJ}$$

Table 6.2. Typical Free Energy Functions (Referred to 298 K)

	$-(G_T^0 - H_{298}^0)/T$ (J/K mol)			$H_{298}^0 - H_0^0$	ΔHf_{298}^0
	298.15 K	500 K	1000 K	(kJ/mol)	(kJ/mol)
$H_2(g)$	130.59	133.89	145.44	8.468	0
$I_2(g)$	260.58	264.81	279.57	10.117	62.43
$HI(g)$	206.48	209.83	221.67	8.657	25.94

and

$$\Delta G^0_{500} = 2(28.0) - 0 - 65.10 \\ - \frac{500}{1000}[2(192.42) - 116.94 - 244.60] = -20.8 \text{ kJ}$$

Notice the division by 1000 to convert J to kJ. Similarly, following the procedure of (6.23) and using the data of Table 6.2 gives

$$\Delta G^0_{298} = 2(25.94) - 0 - 62.43 \\ - \frac{298.15}{1000}[2(206.48) - 130.59 - 260.58] = -17.05 \text{ kJ}$$

and

$$\Delta G^0_{500} = 2(25.94) - 0 - 62.43 \\ - \frac{500}{1000}[2(209.83) - 133.89 - 264.81] = -21.03 \text{ kJ}$$

It may be observed further that, following (6.24), ΔS^0_{298} for the above reaction is, from Table 6.2, 2(206.48) − 130.59 − 260.58 = 21.79 J/K mol. The figure tabulated for the standard entropy of $H_2(g)$ at 298 K has already been referred to in Problem 5.10.

6.7. SUMMARY

In this and the preceding chapter has been shown the method of calculating thermodynamic functions from molecular properties: molecular mass, dimensions, frequencies, symmetry, etc.—the "nuts and bolts" of statistical thermodynamics. One of the most impressive things that should be

fully appreciated is the remarkable agreement between these values and those determined by classical methods, particularly for heat capacity and entropy data. Of the three major kinds of contributions—translational, rotational, and vibrational—classical theory of equipartition, while successful in the first two, was a failure in the third. On the other hand the statistical approach yields the correct, that is, the experimental, result in all cases where discrepancies cannot be accounted for by special explanations such as the presence of residual entropy or nuclear spin isomerism. Where the molecular properties, especially the assignment of frequencies to the various vibrational modes and the magnitudes of the barriers to free rotation, are firmly established, the statistical result is, in fact, likely to be more accurate than the experimental.

PROBLEMS

6.1. Monatomic sodium vapor, $M = 22.99$ g/mol, has an experimental standard entropy (corrected to ideal behavior) at 298.15 K of 153.35 J/K mol. Show how this value is correctly reproduced by statistical thermodynamic calculations.

***6.2.** (a) Nitric oxide, NO, has a doubly degenerate ground electronic level, and a doubly degenerate first excited electronic level which is 12,100 m^{-1} above it. Write the expression for q_e as a function of T and evaluate it for $T = 298.15$ K.

(b) Calculate the electronic contribution (i) to the molar energy and (ii) to the molar entropy at the same temperature.

6.3. Account for the difference between the statistical and third law entropies for CH_3D(g) at its normal boiling point, which are 165.2 and 153.6 J/K mol, respectively.

***6.4.** Find the mole fraction of para-hydrogen in equilibrium hydrogen at 80 K using the value of 87.5 K for the characteristic rotational temperature.

6.5. Calculate the rotational heat capacity for (a) para $H_2(g)$ and (b) "normal" $H_2(g)$ at 150 K, taking θ_r to be 87.5 K.

6.6. (a) For D_2 the ortho isomer has even J values and six nuclear quantum states, and the para isomer has odd J values and three nuclear quantum states. Write the expression for the combined nuclear and rotational partition function in terms of J, θ_r and T, analogous to (6.6).

(b) What should be the mole fraction of ortho-D_2 in equilibrium D_2 (i) at 0 K and (ii) at temperatures high enough for (5.38) to be valid?

***6.7.** For *m*-xylene(g) the vibrational entropy is 50.08 J/K mol at 298.15 K. Given that $M = 106$ g/mol, that I_A, I_B, and I_C are 2.33×10^{-45}, 4.75×10^{-45}, and 6.99×10^{-45} kg m^2, and that I_{ir} for each methyl group is 5.23×10^{-47} kg m^2, compute S^0_{298} and compare with the third law entropy of 358.2 J/K mol. Assume that both methyls rotate freely and that $I_AI_BI_C$ is unaffected by this.

6.8. Find σ_{csn} for (i) *o*-xylene, (ii) mesitylene (1,3,5-trimethylbenzene), (iii) aniline, (iv) 2,2-dimethylpropane.

6.9. (a) Find C_P^0 for $CH_3NO_2(g)$ at 373.15 K from the following data and compare with the measured value of 67.4 $\pm$ 0.8 J/K mol. Atomic weights: H = 1.0, C = 12.0, N = 14.0, O = 16.0 g/mol. $M = 61.04$ g/mol. Bond lengths: C–H = 109, C–N = 146, N–O = 121 pm. Bond angles: O–N–O = 127°, H–C–N = H–C–H = 109.5°. Assume free rotation about the C–N bond. Fundamental vibration frequencies, as wave numbers ($m \times 10^{-2}$): 476, 599, 647, 921, 1097, 1153, 1384, 1413, 1449, 1488, 1582, 2965, 3048 (doubly degenerate).

Table 6.3. Properties of H_2, I_2, and HI Molecules

	M (g/mol)	θ_r (K)	θ_v (K)
H_2	2.016	87.5	5986
I_2	253.81	0.0538	306.8
HI	127.91	9.43	3209

(b) Find S^0_{298} and compare it with the third law value of 275.0 ± 0.8 J/K mol.

***6.10.** From the data given in Table 6.3 and the answer to Problem 5.10(b) calculate the standard free energy function (reference state 0 K) for H_2(g), I_2(g), and HI(g) at 298.15 K and compare with the values given in Table 6.1.

***6.11.** (a) Spectroscopic studies show that D_0, the energy needed to dissociate a diatomic molecule in its lowest vibrational energy level ($v = 0$) into the separate atoms (compare Figure 5.4), is 7.171, 2.470, and 4.896 $\times 10^{-19}$ J for H_2, I_2, and HI, respectively. (Spectroscopists quote these values in electron volts or in cm^{-1}). Find the enthalpy of formation of HI(g) per mole from H_2(g) and I_2(g) at 0 K.

(b) Why does this value differ from 28.0 kJ/mol given in Table 6.1?

6.12. Distinguish clearly between the symmetry number of the rigid molecule CH_3D and the value for W_{tot} at 0 K used to account for its residual entropy in Problem 6.3.

Chapter Seven

CHEMICAL EQUILIBRIUM FOR IDEAL GASES

7.1. INTRODUCTION

The treatment of ideal gases thus far has referred essentially to the determination of the thermodynamic functions for pure substances. In considering chemical equilibrium, however, we are dealing with homogeneous mixtures of gases in which two chemical changes, one the reverse of the other, are occurring at equal rates at a constant total pressure and temperature. One must enquire to what extent the expressions deduced in the earlier chapters are applicable to ideal gases in ideal gas mixtures. Without deriving the result rigorously we will say immediately that the results are applicable without modification. This is because, in a mixture of ideal gases, each gas is independent of the others and behaves as if they were not present, apart from any chemical interactions among them. Just as the molar thermodynamic properties of a pure ideal gas depend only on its pressure and temperature, so also do those of an ideal gas in an ideal gas mixture depend only on its partial pressure and temperature. This is true for E, H, S, G, and A. Moreover, E and H are even independent of pressure. It follows that the values for E^0, H^0, S^0, G^0, and A^0 deduced earlier for pure ideal gases apply to these gases in ideal gas mixtures when they are individually at a *partial* pressure of one atmosphere. For the general chemical equilibrium,

$$aA(g) + bB(g) + \cdots \rightleftharpoons mM(g) + nN(g) + \cdots$$

therefore, ΔG^0, computed using (6.21) or (6.23), is valid not only for the several substances each at one atmosphere and in separate vessels but also for a mixture of the several substances with each one at a partial pressure of one atmosphere.

7.2. DETERMINATION OF EQUILIBRIUM CONSTANTS

For the general chemical equilibrium given above a knowledge of ΔG^0, evaluated as shown in the previous chapter, gives immediate access to the equilibrium constant, K_p, through the well-known relation

$$\Delta G^0 = -RT \ln K_p \tag{7.1}$$

where

$$K_p = \frac{p_M^m p_N^n \cdots}{p_A^a p_B^b \cdots} \tag{7.2}$$

in which the p's are the partial pressures at equilibrium. [It should be noted from the usual derivation of (7.2) given in all classical thermodynamics texts that each p is really p/p^0, the ratio of the partial pressure to the pressure in the standard state. If, as is customary, the standard state is 1 atm, all the p's are in atm and all the p^0's are unity. The latter do not therefore appear in (7.2). Furthermore, since each p/p^0 is dimensionless, so also is K_p.] Thus, for example, a knowledge of the standard free energy functions, referred to 0 K, for the various species in chemical equilibrium, all of which can be determined by the statistical approach already described in the previous chapter, and a knowledge of ΔH_0^0 can readily lead to ΔG^0 and thence to K_p. It is worth emphasizing that in (6.21) and (6.22) the quantities with a subscript zero refer to the ground state for every species. When the free energy functions are determined by statistical thermodynamics the equilibrium constant derived therefrom is a statistical thermodynamic result.

EXAMPLE 7.1. Using the data in Table 6.1 determine K_p at 298 K for $H_2(g) + I_2(g) \rightleftharpoons 2HI(g)$.

ANSWER. As already shown in Section 6.6, $\Delta G^0_{298} = -16.8$ kJ, using (6.21). From (7.1) $K_p = e^{(-\Delta G0/RT)} = \exp[(16{,}800/8.314 \times 298.15)] = 8.8 \times 10^2$.

There is an alternative but equivalent approach to the evaluation of K_p which is more laborious but more instructive. We begin by applying (4.20) to the standard state, namely,

$$G^0 \text{ (per mole)} = -kNT \ln (q^0/L) = -RT \ln (q^0/L) \quad (7.3)$$

To emphasize that this is the value above the ground state we write this as

$$G^0 \text{ (per mole)} - H_0^0 \text{ (per mole)} = -RT \ln (q^0/L) \quad (7.4)$$

recalling that for ideal gases $H_0^0 = G_0^0$ (see the last paragraph of Section 5.6). It follows that

$$G^0 \text{ (per mole)} = H_0^0 \text{ (per mole)} - RT \ln (q^0/L) \quad (7.5)$$

If we now apply this result to every species in the general equilibrium $aA(g) + bB(g) + \cdots \rightleftharpoons mM(g) + nN(g) + \cdots$ and subtract the sum of the values of G^0 for the reactants from the sum of the values for the products we obtain

$$\begin{aligned} mG_M^0 + nG_N^0 + \cdots - aG_A^0 - bG_B^0 - \cdots \\ = mH_{0M}^0 + nH_{0N}^0 + \cdots - aH_{0A}^0 - bH_{0B}^0 - \cdots \\ -RT \ln \{[(q_M/L)^m(q_N/L)^n \cdots]/[(q_A/L)^a(q_B/L)^b \cdots]\} \end{aligned}$$

or

$$\Delta G^0 = \Delta H_0^0 - RT \ln \frac{(q_M^0/L)^m (q_N^0/L)^n \cdots}{(q_A^0/L)^a (q_B^0/L)^b \cdots} \tag{7.6}$$

If (7.6) is divided by $-RT$ the result can be equated to $\ln K_p$ according to (7.1) giving

$$\ln K_p = \ln \frac{(q_M^0/L)^m (q_N^0/L)^n \cdots}{(q_A^0/L)^a (q_B^0/L)^b \cdots} - \frac{\Delta H_0^0}{RT} \tag{7.7}$$

Finally, taking antilogarithms yields

$$K_p = \frac{(q_M^0/L)^m (q_N^0/L)^n \cdots}{(q_A^0/L)^a (q_B^0/L)^b \cdots} e^{-\Delta H_0^0/RT} \tag{7.8}$$

Every q in (7.8) is, of course, a total molecular partition function and is itself a product of q_t, q_r, q_v, q_e, etc. as already stated in (4.26).

It is evident from (7.8) that K_p is determined by two factors, one involving the q's for every species involved in the equilibrium and the other, an exponential factor, involving ΔH_0^0 ($= \Delta E_0^0$), the energy change for the reaction with every species in its ground state. Thus K_p can be calculated from the partition functions and the ground state energy difference between reactants and products, all species being in their standard states. The evaluation of equilibrium constants using (7.8) is, as already stated, equivalent to its evaluation from standard free energy functions through ΔG^0, but the extensive arithmetic involved in computing and using partition functions appears explicitly instead of being hidden in the free energy functions.

One important point needs to be emphasized: the choice of energy zeros must be consistent. If, for each of the vibrational modes of the various molecular species, the ground state ($v = 0$) is chosen as the energy zero, Eq. (5.70), then ΔH_0^0 must be the energy change in going from reactants in the ground vibrational state to products in their ground vibrational state. If, on the other hand, for each of the vibrational modes of the various species, the hypothetical motionless state (bottom of the potential energy well) is chosen as the energy zero, Eq. (5.69), then ΔH_0^0 must be the energy change in going from reactants in the vibrationless states to the products in their vibrationless states.

Since it may be helpful, for the calculations of this and subsequent sections, to compute directly the translational quantities q_t^0 and q_t^0/L, we assemble here convenient expressions for this purpose. We return to (5.8) and apply it at a pressure of 1 atm (P^0). As we are using it to express *molar* thermodynamic properties, for example, molar free energy, V refers to the volume occupied by a mole of the ideal gas at 1 atm pressure. Hence V may be replaced by RT/P^0. We have, therefore,

$$q_t^0 = \frac{[2\pi(M/1000L)kT]^{3/2}RT}{h^3(101{,}325)}$$

$$= \frac{[2\pi(1/1000L) \times 1.38066 \times 10^{-23}]^{3/2} \times 8.31441}{(6.626176 \times 10^{-34})^3 \times 101{,}325} M^{3/2}T^{5/2}$$

or

$$q_t^0 = 1.5421 \times 10^{22} M^{3/2} T^{5/2} \tag{7.9}$$

with M in g/mol. For $T = 298.15$ K this becomes

$$q_t^0 = 2.3670 \times 10^{28} M^{3/2} \tag{7.10}$$

From these it follows immediately that

$$q_t^0/L = 2.5607 \times 10^{-2} M^{3/2} T^{5/2} \tag{7.11}$$

and, for $T = 298.15$ K,

$$(q_t^0/L) = 3.9306 \times 10^4 M^{3/2} \tag{7.12}$$

Similarly, q_r for linear molecules, given by (5.38), can be rewritten

$$q_r = 2.4829 \times 10^{45} IT/\sigma \tag{7.13}$$

with I in kg m^2. For $T = 298.15$ this becomes

$$q_r = 7.4026 \times 10^{47} I/\sigma \tag{7.14}$$

For nonlinear molecules (5.48) can be rewritten

$$q_r = 2.1928 \times 10^{68} \frac{(I_A I_B I_C)^{1/2}}{\sigma} T^{3/2} \tag{7.15}$$

which, for $T = 298.15$ K, becomes

$$q_r = 1.1289 \times 10^{72} \frac{(I_A I_B I_C)^{1/2}}{\sigma} \tag{7.16}$$

Before applying (7.8) it may be rearranged as follows for convenience. Replacing every q^0 by $q_t^0 q_r q_v \cdots$ and regrouping gives

$$K_p = \left[\frac{q_{tM}^{0m} q_{tN}^{0n} \cdots}{q_{tA}^{0a} q_{tB}^{0b} \cdots}\right] \left[\frac{q_{rM}^{m} q_{rN}^{n} \cdots}{q_{rA}^{a} q_{rB}^{b} \cdots}\right] \left[\frac{q_{vM}^{m} q_{vN}^{n} \cdots}{q_{vA}^{a} q_{vB}^{b} \cdots}\right] \left[\frac{\cdots}{\cdots}\right] \times L^{(a+b+\cdots)-(m+n+\cdots)} \times e^{-\Delta H_0^0/RT} \tag{7.17}$$

In the special, but common, situation in which $a + b + \cdots = m + n + \cdots$ the first factor in square brackets reduces to

$$\text{translational factor} = (M_M^m M_N^n \cdots / M_A^a M_B^b \cdots)^{3/2} \quad (7.18)$$

and the factor in L disappears. The second (rotational) factor likewise can often be reduced, for example, if all the species are diatomic, but such simplifications must be made with care.

EXAMPLE 7.2. Calculate K_P at 298 K for $H_2(g) + I_2(g) \rightleftharpoons 2HI(g)$ using (7.8) and the data given in Table 7.1, and show that one arrives at the same result as found in Example 7.1. For H_2 use the value of q_r given in the answer to Problem 5.10(c), namely, 1.881.

ANSWER. Since there are two molecules on each side of the equation ($m + n + \cdots = a + b + \cdots$) we may use a simplification mentioned above. The translational factor in (7.17) reduces, according to (7.18), to $127.91^3/(2.016 \times 253.81)^{3/2} = 180.81$. Because of the special treatment afforded H_2, the rotational factor is best obtained as follows: for H_2, $q_r = 1.881$ (given); for I_2, by (5.38), $q_r = 298.15/2(0.0538) = 2771$; for HI, similarly, $q_r = 298.15/9.43 = 31.62$. The rotational factor is therefore $31.62^2/1.881 \times 2771 = 0.1918$. For H_2 and HI, θ_v is large enough to give $q_v = 1.000$ [Eq. (5.70)]; for I_2, $q_v = (1 - e^{-306.8/298.15})^{-1} = 1.556$. The vibrational factor is therefore $1.000^2/1.000 \times 1.556 = 0.6426$. The factor in L disappears.

To find ΔH_0^0 we note that 7.171×10^{-19} and 2.470×10^{-19} J are needed to separate the atoms of H_2 and of I_2, respectively, and that $2(4.896 \times 10^{-19})$ J is recovered when the atoms are recom-

Table 7.1. Properties of H_2, I_2, and HI Molecules

	$M/\text{g mol}^{-1}$	θ_r/K	θ_v/K	$D_0/10^{-19}$ J
$H_2(g)$	2.016	87.5	5986	7.171
$I_2(g)$	253.81	0.0538	306.8	2.470
HI(g)	127.91	9.43	3209	4.896

bined to give 2HI. ΔH_0^0 is thus $[7.171 + 2.470 - 2(4.896)]10^{-19}$ or -1.51×10^{-20} J for the formation of two molecules of HI. Therefore for the formation of two moles of HI, $\Delta H_0^0 = 1.51 \times 10^{-20}L = -9.09 \times 10^3$ J.

Finally, making these substitutions in (7.17) yields $K_P = 180.81 \times 0.1918 \times 0.6426 \times \exp[-(-9.09 \times 10^3/8.31 \times 298)] = 8.8 \times 10^2$ as found in Example 7.1.

EXAMPLE 7.3. Determine, once again, the value of K_P, for $H_2(g) + I_2(g) \rightleftharpoons 2HI(g)$ at 298 K using (7.8) and the data provided in Example 7.2, but taking the vibrationless states for all the vibrational energy zeros.

ANSWER. The translational and rotational factors in (7.8) are the same as given in the Answer to Example 7.2, namely, 180.81 and 0.1918, respectively. For the vibrational partition functions, however, (5.69) is used. Therefore, for H_2,

$$q_v = e^{-5986/2\times 298.15}(1 - e^{-5986/298.15})^{-1} = 4.37 \times 10^{-5}$$

for I_2, similarly, $q_v = 0.930$; and for HI, $q_v = 4.60 \times 10^{-3}$. The vibrational factor is therefore $(4.60 \times 10^{-3})^2/4.37 \times 10^{-5} \times 0.930 = 0.521$. The value of ΔH_0^0 will now be the energy change with all species at the bottom of the potential energy well, that is, with vibrational energies $\frac{1}{2}h\nu_0$ lower than the v = 0 level (see Figure 5.4). For each species this energy will be $\frac{1}{2}h\nu_0 + D_0$ lower than that for the separated atoms. Since $\theta_v = h\nu_0/k$, $\frac{1}{2}h\nu_0 = k\theta_v/2$ or $R\theta_v/2$ per mole, so for one mole $\frac{1}{2}h\nu_0 + D_0$ becomes $R\theta_v/2 + LD_0$. Insertion of the data provided gives values of 456.73, 150.02, and 308.18 kJ/mol for H_2, I_2, and HI, respectively. Therefore $\Delta H_0^0 = 456.73 + 150.02 - 2\ (308.18) = -9.61$kJ. Thus K_P if given by $180.81 \times 0.1918 \times 0.521 \times \exp[-(-9.61 \times 10^3/8.31 \times 298)] = 8.7 \times 10^2$ as found in Examples 7.1 and 7.2.

Another type of equilibrium to which (7.8) applies is the equilibrium between a monatomic metal atom and its positive ion as exemplified by

$$Na(g) \rightleftharpoons Na^+(g) + e^-$$

The resulting free electrons can be treated as ideal gases, as can the cations formed. The one valence electron of Na(g) can have two spin states, as can the free electron. This must be recognized by the use of the correct degeneracies. The ground electronic state of Na(g), for example, is $J = ½$, where J is the total angular momentum quantum number of the electrons in the atom and, since the multiplicity is given by $2J + 1$, $g_0 = 2$. Furthermore, since the masses of the atom and its ion are virtually equal, they will cancel when K_p is evaluated. The value of ΔH_0^0 corresponds to the ionization energy, usually quoted in electron volts (eV). (1 eV = 1.602×10^{-19} J.) There will, of course, be no rotational or vibrational contributions for monatomic atoms or ions, or free electrons.

EXAMPLE 7.4. Determine K_P for $Na(g) \rightleftharpoons Na^+(g) + e^-$ at 5000 K, given that the atomic weight of Na is 22.99 g/mol, the mass of an electron is 9.111×10^{-31} kg, and the total angular momentum quantum number of the ground electronic state of Na(g) is $J = ½$. The first ionization potential (energy) of Na is 5.14 eV. Higher excited electronic states of the sodium are inaccessible. The free electrons may be treated as ideal gas molecules. (1 eV = 1.602×10^{-19} J.)

ANSWER. According to (7.8)

$$K_P = (q^0_{Na+}/L)(q^0_{e-}\ /L)/(q^0_{Na}/L)e^{-\Delta H_0^0/RT}$$

Since the masses and temperatures of the Na and Na^+ are the same, the translational q^0's [see (5.8)] will cancel, as will the rotational and vibrational q's, which are all unity. For Na(g), $q_e = (2J + 1)e^{-0} = 2$, but for Na^+(g), $q_e = 1$ (no unpaired electrons). Thus $(q^0_{Na+}/L)/(q^0_{Na}/L) = ½$. For the free electrons $q^0/L = (q^0_t/L)q_e$, and q^0_t/L may be evaluated using (7.11), where, however, M must be in g/mol or $9.111 \times 10^{-28}L = 5.49 \times 10^{-4}$ g/mol. Moreover, q_e for the electrons (doubly degenerate) $= 2e^{-0} = 2$. Therefore $q^0/L = 2.5607 \times 10^{-2}(5.49 \times 10^{-4})^{3/2}(5000)^{5/2}2 = 1165$ for the electrons. Finally, since the ionization potential is $5.14 \times 1.602 \times 10^{-19}L =$

4.96×10^5 J/mol, we have $K_P = \frac{1}{2} \times 1165 \exp[-4.96 \times 10^5/8.314(5000)] = 3.83 \times 10^{-3}$.

7.3. STATISTICAL INTERPRETATION OF EQUILIBRIUM CONSTANTS

The approach to the evaluation of equilibrium constants through (7.8) casts light on their statistical basis as follows. We note, first, that the right-hand side of (7.8) is the product of two principal factors, one containing the partition functions of all the species directly involved in the reaction, and the other, an exponential factor, containing an energy difference. It may be observed, in passing, that classical thermodynamics leads to a comparable result: by equating the right hand sides of (7.1) and the familiar relation

$$\Delta G^0 = \Delta H^0 - T\Delta S^0 \tag{7.19}$$

dividing through by $-RT$, and taking antilogarithms, one obtains

$$K_P = e^{\Delta S^0/R} e^{-\Delta H^0/RT} \tag{7.20}$$

—again the product of two factors, one of which contains an energy difference. However, classical thermodynamics offers no physical interpretation of (7.20) whereas statistical thermodynamics does indeed offer one for (7.8).

Let us consider, for simplicity, a chemical equilibrium between the two isomeric butanes, n-C_4H_{10} and i-C_4H_{10}, which will be abbreviated by A and B, respectively:

$$n\text{-}C_4H_{10}(g) \rightleftharpoons i\text{-}C_4H_{10}(g)$$

Application of (7.8) to this equilibrium yields

$$K_p = \frac{p_B}{p_A} = \frac{q_B^0}{q_A^0} e^{-\Delta H_0^0/RT} \tag{7.21}$$

Imagine that, by bringing together 4 moles of carbon atoms and 10 moles of hydrogen atoms at a given temperature, only the two species, A and B, can form. (There are, of course, other conceivable species which might result, but we shall for simplicity neglect such possibilities.) Any molecule of species A has a very extensive set of energy levels available to it, and any molecule of species B also has a very extensive, but different, set of energy levels available to it. Any set of four carbon and ten hydrogen atoms thus "has a choice" as to whether it will form A molecules and occupy one of the A energy levels or form B molecules and occupy one of the B energy levels. The "choice" is not a permanent one for there are continual exchanges from one energy level to another without changing A to B or B to A, as well as continual interconversions of A and B. At equilibrium, however, in spite of this internal activity the ratio of the partial pressures and therefore the ratio of the number of molecules of B to A remains constant at the value given by K_p. Equation (7.21) tells us that this ratio depends on (i) q_B^0/q_A^0 and (ii) $e^{-\Delta H_0^0}/RT$.

The influence of these two factors may be discussed profitably by referring to Figure 7.1, which is a schematic diagram depicting the energy levels in A and B at room temperature. Because A is a less rigid (more "floppy") molecule than B there are more available energy levels for it than for B in a given energy range above ground. The spacing of the levels in A is thus smaller than the spacing in B. This fact, in itself, causes q_A^0 to be larger than q_B^0 (compare Section 1.4) and favors the formation of A molecules. For this reason the factor q_B^0/q_A^0 is

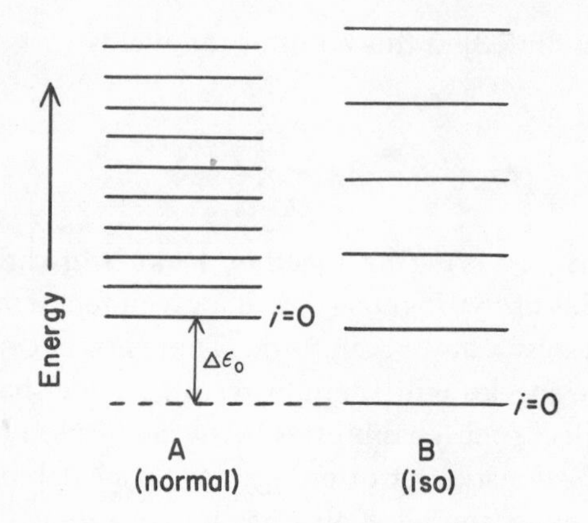

Figure 7.1. Comparison of energy levels in normal and isobutane (schematic).

greater than unity in this particular system. However, as shown in Figure 7.1, the energy of the ground state of B is lower than that of A by an amount $\Delta\epsilon_0$ per molecule, and this fact, by itself, "attracts" molecules into the B structure since, according to the Boltzmann distribution, Eq. (1.15), levels of lower energy have the greater populations. This effect is contained in the factor $e^{-L\Delta\epsilon_0^0/RT} = e^{-\Delta H_0^0/RT}$, since $\Delta H_0^0 = L\Delta\epsilon_0^0$. Thus in this system we have two opposing drives, the reduced spacing of the A levels favoring the formation of A and lower ground level energy of B favoring the formation of B. When the actual numerical values are inserted it is found that, at 298 K, $q_B/q_A = 0.160$ and, since $\Delta H_0^0 = -6.82$ kJ, $e^{-\Delta H_0^0/RT} = 15.7$, so that $K_p = 0.160 \times 15.7 = 2.50$. Since $K_p > 1$ there is more of the iso form than of the normal form present at equilibrium at 298 K. It should be added, however, that the exponential factor is more temperature-sensitive than the partition function factor, so that at sufficiently high temperatures the former approaches unity and gives $K_p < 1$, with a consequent preponderance of the normal form. This corresponds to saying in more classical language that in (7.19) the $T\Delta S^0$ term dominates the ΔH^0 term at high temperatures. The sign of $T\Delta S^0$ then determines the sign of ΔG^0 and thus whether K_p is less than or greater than unity.

7.4. ISOTOPE EXCHANGE EQUILIBRIA

An interesting application of (7.8) is in the area of isotope exchange. We shall see that the position of equilibrium is, to a great extent, controlled by symmetry considerations.

Consider, for example, the equilibrium

$$H_2(g) + D_2(g) \rightleftharpoons 2HD(g)$$

at 400 K. In applying (7.8) to this we shall, in principle, need to compute q_t^0/L, q_r, and q_v for each of the three species. When these are grouped so that all the translational factors are together, all the rotation factors together, and so on, we will have

$$K_P = \left[\frac{(q_{t\mathrm{HD}}^0/L)^2}{(q_{t\mathrm{H_2}}^0/L)(q_{t\mathrm{D_2}}^0/L)}\right]\left[\frac{q_{r\mathrm{HD}}^2}{(q_{r\mathrm{H_2}})(q_{r\mathrm{D_2}})}\right]\left[\frac{q_{v\mathrm{HD}}^2}{(q_{v\mathrm{H_2}})(q_{v\mathrm{D_2}})}\right]e^{-\Delta H_0^0/RT}$$

The first factor in square brackets, however, reduces through cancellation to $M_{\mathrm{HD}}^3/M_{\mathrm{H_2}}^{3/2}M_{\mathrm{D_2}}^{3/2}$ according to (7.18). The second factor, likewise, reduces to $(I_{\mathrm{HD}}/\sigma_{\mathrm{HD}})^2/(I_{\mathrm{H_2}}/\sigma_{\mathrm{H_2}})(I_{\mathrm{D_2}}/\sigma_{\mathrm{D_2}})$ on applying (5.38), provided T is large enough for it to be valid for diatomic hydrogen. The third factor [see (5.68)] and the exponential factor do not reduce by cancellation. Thus we have

$$K_p = \left[\frac{M_{\mathrm{HD}}^3}{M_{\mathrm{H2}}^{3/2} M_{\mathrm{D2}}^{3/2}}\right]\left[\frac{I_{\mathrm{HD}}^2}{I_{\mathrm{H2}} I_{\mathrm{D2}}}\right]\left[\frac{\sigma_{\mathrm{H2}}\sigma_{\mathrm{D2}}}{\sigma_{\mathrm{HD}}^2}\right] \times \left[\frac{[(1 - e^{-\Theta_{v\mathrm{HD}}/T})^{-1}]^2}{(1 - e^{-\Theta_{v\mathrm{H2}}/T})^{-1}(1 - e^{-\theta_{v\mathrm{D2}}/T})^{-1}}\right] e^{-\Delta H_0^0/RT}$$

The data given in Table 7.2 are available. We have, therefore, since $\Delta H_0^0 = [431.8 + 439.2 - 2(435.2)]1000 = 6 \times 10^2$ J,

$$\begin{aligned} K_p &= \frac{3.0219^3}{(2.0156)^{3/2}(4.0282)^{3/2}} \times \frac{(6.1303 \times 10^{-48})^2}{(4.6030 \times 10^{-48})(9.1955 \times 10^{-48})} \\ &\quad \times \frac{2 \times 2}{1} \times \frac{[(1 - e^{-5226/400.0})^{-1}]^2}{(1 - e^{-5986/400.0})^{-1}(1 - e^{-4308/400.0})^{-1}} \\ &\quad \times \exp(-6 \times 10^2/8.314 \times 400.0) \\ &= 1.1928 \times 0.8879 \times 4 \times 1.0000 \times 0.83 \\ &= 3.52. \end{aligned}$$

It will be evident that, in general, the factors containing the M's and I's will reduce to quantities not far from unity, as in this example. The same is true for the factor containing θ_v, and for the factor containing ΔH_0^0, since the latter will be small. Therefore K_p would be near unity were it not for the factor with the symmetry numbers. It may be said, then, that the position of equilibrium is controlled chiefly by symmetry considerations, favoring the side with less symmetry. The

Table 7.2. Properties of H_2, D_2, and HD Molecules

	M/g mol^{-1}	I/kg m$^2 \times 10^{-48}$	θ_v/K	D_0/kJ mol^{-1}
H_2	2.0156	4.6030	5986	431.8
D_2	4.0282	9.1955	4308	439.2
HD	3.0219	6.1303	5226	435.2

influence of symmetry on the position of equilibrium is an important matter, and will be considered in the next section.

The alternative methods of finding K_P by means of (7.8), illustrated in Examples 7.2 and 7.3, can be applied to isotope exchange equilibria, but a simplification emerges: Consider, again, the equilibrium $H_2(g) + D_2(g) \rightleftharpoons 2HD(g)$ just discussed. The translational and rotational factors in the second method remain as given above. If, as was done in the Answer to Example 7.3, the hypothetical vibrationless species are chosen as the energy zeros, we have, for H_2

$$q_v = e^{-5986/2\times400}(1 - e^{-5986/400})^{-1} = 5.628 \times 10^{-4}$$

according to (5.67). Analogously, q_v is found to be 4.58×10^{-3} for D_2 and 1.46×10^{-3} for HD. The vibrational partition function factor in (7.8) is thus $(1.46 \times 10^{-3})^2/5.63 \times 10^{-4} \times 4.58 \times 10^{-3} = 0.827$. Because, however, the potential energy curves (Figure 5.4) for all three species may be assumed to be the same (since the species differ only isotopically), $\frac{1}{2}h\nu_0 + D_0 = D_e$ is the same for all of them. It follows that $\Delta H_0^0 = 0$ on this basis. Thus $K_P = 1.1928 \times 0.8879 \times 4 \times 0.827 = 3.50$, as found above.

EXAMPLE 7.5. A mixture of light and heavy water vapor equilibrates according to $H_2{}^{16}O(g) + D_2{}^{16}O(g) \rightleftharpoons 2H^{16}OD(g)$. On the basis of symmetry only, estimate the equilibrium constant at 800 K.

ANSWER. By an extension of the ideas described above to polyatomic molecules we may assume that K_P will be close to unity except for the factor $(\sigma_{H_2O})(\sigma_{D_2O})/\sigma_{HOD}^2$ which will result from the use of (5.49) for q_r for each species. Since $\sigma_{H_20} = \sigma_{D_20} = 2$ and $\sigma_{HOD} = 1$, this factor has the value 4. We may predict, therefore, that K_P will be about 4.

7.5. ESTIMATION OF POSITION OF EQUILIBRIUM

It is sometimes possible to estimate the position of equilibrium by a qualitative comparison of the energies and entropies of reactants and products. In terms of (7.17) this means estimating the effects of the translational, rotational, vibrational, etc. factors (in square brackets) and of the final exponential factor. In equilibria between pairs of isomers, for example, the translational factor in (7.17) will be unity, of course, and if the isomers are closely similar, one can expect that the products of their moments of inertia will also be similar. This will imply that the rotational factor in (7.17) will be unity also, except for possibly different symmetry numbers. The vibrational factor may also not be far from unity, since the same kinds of bonds will, in general, be found in both isomers. The value of ΔH_0^0, likewise, could be close to zero unless some effects caused by such things as steric hindrance, are present. The existence of the latter may immediately determine at least the sign of ΔH_0^0.

Consider the equilibrium *m*-xylene(g) $\rightleftharpoons$ *p*-xylene(g), for example. Assuming, as just described, that the translational and vibrational factors are near unity, and that the rotational factor, apart from the symmetries, is unity also, we are led to consider the latter. For *m*-xylene $\sigma = 2$ but for *p*-xylene $\sigma = 4$. Thus the rotational factor, including symmetry, would be 2/4 or 0.5. (If we include a factor for internal rotation this ratio would be 18/36, which is still 0.5.) Thus this factor favors the *m*-xylene. A consideration of ΔH_0^0 leads one to expect this quantity to be nearly zero, there being little difference between the isomers energetically. The exponential factor in (7.17) is thus expected to be near unity. Thus the only significant factor which is not near unity is the symmetry ratio. One would then be led to predict that K_p is about 0.5, close to the value determined more accurately, namely, 0.40 at 298 K. The equilibrium is thus toward the *meta* isomer.

EXAMPLE 7.6. Predict the position of equilibrium between *m*-xylene(g) and *o*-xylene(g).

ANSWER. One can make the same assumptions as were made for the *m*- and *p*-xylenes above, recognizing that the internal rotation of the methyl groups in *o*-xylene will not be as free as it is in *m*-xylene. Since σ_{csn} is 18 for both *m*- and *o*-xylenes any thermodynamic differences may be attributed largely to a difference in the freedom of the internal rotations. As stated in Section 6.5, an internal rotation which is hindered increases the free energy over what it would be for a free rotation. The free energy of the *ortho* form should thus be larger than that of the *meta*, so that, for *m*-xylene $\rightleftharpoons$ *o*-xylene, the position of equilibrium should be to the left, and $K_P < 1$ is predicted. (The value by more quantitative estimate is 0.27.)

Admittedly, the examples just discussed are fairly clear-cut but many situations can arise where a decision is more difficult to reach. For example, when symmetry considerations lead to one conclusion but energy considerations lead to the opposite conclusion, it may be impossible to come even to a qualitative decision.

7.6. CONCLUDING COMMENTS

It should be evident, by the time the reader has assimilated the material in this book, that the statistical approach to the evaluation of thermodynamic quantities is, at least in the areas of atomic solids and ideal gases, a powerful and enlightening tool. With a knowledge of the most elementary properties of molecules, their molecular weights, bond lengths, bond angles, symmetry numbers, and energies, it permits the computation of all the thermodynamic properties of pure gaseous substances, and the position of equilibrium in mixtures of interacting gases. The results agree impressively with those found by classical thermodynamic methods and, where differences between the two methods are found, it is usually possible to account for them. In such cases it is often the statistical result which is the more accurate of the two.

The extension of the concepts outlined in this text to nonideal gases and liquids is an obvious next step. It is, however, a difficult one, and the development of these aspects is in its infancy compared with that of ideal gases. This is because of the horrendous difficulties that ensue when the potential energy of molecules in close proximity to others must be taken into account. It is hoped that the study of this book will induce the reader to appreciate the power and beauty of the subject and to investigate its further development. As a first step he or she should read more advanced texts which deal with the subject of ensembles.

PROBLEMS

7.1. (a) Calculate the values of $-(G_T^0 - H_0^0)/T$ for $C_2H_4(g)$ and $C_2H_6(g)$ at 298.15 K from the following information given in Table 7.3. For C_2H_6 assume free rotation about the C–C bond with $I_{ir} = 2.65 \times 10^{-47}$ kg m^2.
(b) The tabulated values of $-(G_{298}^0 - H_0^0)/298.15$ for $C_2H_4(g)$ and $C_2H_6(g)$ are 184.01 and 189.41 J/mol, respectively. To what do you attribute any appreciable differences between these results and those found in (a)?
(c) For $H_2(g)$, $-(G_{298}^0 - H_0^0)/298.15$ is 102.17 J/mol. Combine this with the data given in (b) to compute Δ

Table 7.3. Properties of C_2H_4 and C_2H_6 Molecules

	M/g mol^{-1}	I_A	I_B	I_C	Contributing fundamental frequencies (cm^{-1})[a]
		($\times 10^{47}$ kg m^2)			
C_2H_4	28.05	5.7	27.5	33.2	900, 940, 950(2), 1100, 1342, 1444, 1623
C_2H_6	30.07	10.7	40.1	40.4	827(2), 993, 1120(2), 1375, 1380, 1460(2), 1480(2)

[a]Degeneracies in parentheses.

G^0 and K_P at 298 K for $C_2H_4(g) + H_2(g) \rightleftharpoons C_2H_6(g)$. $\Delta H_0^0 = -129.87$ kJ.

***7.2.** Find K_P at 400 K for $CO(g) + H_2O(g) \rightleftharpoons CO_2(g) + H_2(g)$ from the data in Table 7.4.

7.3. Determine K_P for $N_2(g) + O_2(g) \rightleftharpoons 2NO(g)$ at 2000 K. M/g mol^{-1} = 28.0, 32.0, 30.0; θ_r/K = 2.89, 2.08, 2.45; θ_v/K = 3353, 2239, 2699; D_0/kJ mol^{-1} = 941.2, 490.1, 626.1, respectively. O_2 has a triply degenerate ground electronic state and a doubly degenerate excited electronic state lying 1.5733×10^{-19} J above ground. NO has a doubly degenerate ground electronic state and a doubly degenerate excited electronic state lying 2.3838 $\times 10^{-21}$ J above ground.

***7.4.** Calculate K_P at 500 K for $N_2(g) + 3H_2(g) \rightleftharpoons 2NH_3(g)$ with the help of the information given in Table 7.5. The enthalpy of formation of $NH_3(g)$ is $\Delta H_0^0 = -39.2$ kJ/mol.

***7.5.** For the ionization of a gaseous alkali metal atom, M(g) $\rightleftharpoons M^+(g) + e^-$, the ground electronic state of the metal atom is doubly degenerate, as are the electrons. Find a general expression for the equilibrium constant for this dissociation as a function of the temperature, T, and the

Table 7.4. Properties of CO, H_2O, CO_2, and H_2 Molecules

	M (g/mol)	I_A	I_B	I_C	Vib. frequency (cm^{-1})	ΔH^0_{0f} (kJ/mol)
		(kg $m^2 \times 10^{47}$)				
CO(g)	28.01	14.49	—	—	2143.2	−113.81
H_2O(g)	18.01	1.02	1.92	2.94	3652, 1595, 3756	−238.94
CO_2(g)	44.01	71.8	—	—	1314, 2335, 663(2)	−393.17
H_2(g)	2.016	0.460	—	—	4160.2	0

first ionization potential, I, in eV. The electron mass is 9.111×10^{-31} kg. Treat the free electrons as an ideal gas.

7.6. Prove that the expression for K_P given by (7.17) and valid when the ground vibrational states are the energy zeros is valid also when the hypothetical vibrationless molecules are the energy zeros. Assume, for simplicity, that all the species are diatomic molecules and that all the coefficients a, b, m, n, etc., are unity.

7.7. On the basis of the arguments presented in Section 7.3 predict whether the equilibrium

$$CH_3 \cdot CH(CH_3) \cdot CH_2 \cdot CH_3(g) \rightleftharpoons C(CH_3)_4(g)$$

lies to the right or to the left at high temperatures.

Table 7.5. Properties of N_2, H_2, and NH_3 Molecules

	M/g mol^{-1}	θ_r/K	θ_v/K
N_2	28.02	2.89	3353
H_2	2.016	87.5	5986
NH_3	17.03	14.30, 14.30, 9.08	4912(2), 4801, 2342(2),1367

7.8. Using (1.17) for both A and B in the isomeric equilibrium $A(g) \rightleftharpoons B(g)$ and applying the result to the respective ground states of A and B, derive (7.8).

***7.9.** Determine K_P for $^{16}O_2(g) + {}^{18}O_2(g) \rightleftharpoons 2\,{}^{16}O^{18}O(g)$ at 1000 K. Assume that the bond lengths are the same in all three diatomic molecules. The fundamental vibration frequencies for $^{16}O_2$, $^{18}O_2$, and $^{16}O^{18}O$ are 1580, 1490, and 1536 cm^{-1}, respectively. Atomic weights: $^{16}O = 16.0$, $^{18}O = 18.0$.

7.10. (a) Find K_P at 800 K for $H_2O(g) + D_2O(g) \rightleftharpoons 2HOD(g)$ from the data given in Table 7.6. Neglect any vibrational contributions to the free energy.
(b) Compare the value with the result of Example 7.5. To what principal cause do you attribute any difference?

7.11. Determine K_P for the dissociation of KBr(g) into its ions at 700 K. The bond length in KBr(g) is 294 pm and $\Delta H_0^0 = 472$ kJ/mol. Ignore any vibrational excitation. Atomic weights: K = 39, Br = 80. Treat the gaseous ions as ideal gas molecules.

***7.12.** (a) The equilibrium constant, K_P, for $I_2(g) \rightleftharpoons 2I(g)$ is found experimentally to be 1.23 at 1200 K. Determine ΔH_0^0 for this reaction if the bond length in I_2 is 266.6 pm and its fundamental vibration frequency is 213.2 cm^{-1}.

Table 7.6. Properties of H_2O, D_2O, and HOD Molecules

	M/g mol^{-1}	I_A	I_B	I_C	Zero-point energy/kJ mol^{-1}
		(kg m^2 × 10^{47})			
H_2O	18.01	1.02	1.92	2.94	231.92
D_2O	20.03	1.84	3.83	5.67	169.58
HOD	19.02	1.21	3.06	4.27	201.79

Atomic weight of I = 126.9. For the ground electronic state of I(g) J = 3/2. Higher electronic states are not appreciably populated at 1200 K.
(b) The *spectroscopic* dissociation energy of I_2, 239.3 kJ/mol, is larger than the value obtained in (a) because spectroscopic dissociation give two iodine atoms, one of which is in the J = 3/2 (ground) state but the other is in the J = 1/2 (excited) state. How much more energy, in eV, has the J = 1/2 state than the J = 3/2 state?

***7.13.** (a) Calculate the product of the moments of inertia of the methyl radical, CH_3, assuming it to be planar trigonal with bond lengths 110 pm. Atomic weights: H = 1.0, C = 12.0.
(b) Find the free energy function (referred to 0 K) for CH_3 at 298 K using the result in (a). Assume that all the vibrations are in the ground state. Take g_0 = 2 for the ground electronic state, with no excited states occupied.
(c) Repeat (a) and (b) for CH_4 with bond lengths 110 pm. To find $I_A I_B I_C$ use the relation that each moment of inertia is given by (8/3)(m_H in kg/mol)(bond length in meters)^2. Take q_v = 1.000 as for CH_3.
(d) The free energy functions (referred to 0 K) for ^{35}Cl and $H^{35}Cl(g)$ at 298 K are 144.06 and 157.82 J/K mol,

Table 7.7. Properties of *Trans* and *Gauche* 1,2-Dichloroethane

	$I_A I_B I_C / 10^{-135}$ kg^3 m^6	Contributing vibrational frequencies/cm^{-1}
Trans	9.21	223, 300, 709, 754, 1052
Gauche	12.12	265, 411, 654, 677, 1031

respectively. Determine K_P at 298 K for $CH_4(g) + {}^{35}Cl(g) \rightleftharpoons CH_3(g) + H^{35}Cl(g)$, taking ΔH_0^0 to be -5 kJ.

7.14. (a) For benzoic acid (C_6H_5COOH) and 2,4-dinitrophenol ($HO \cdot C_6H_3(NO_2)_2$) the acid dissociation constants, K_a, in water are 6.31×10^{-5} and 8.13×10^{-5}, respectively, at 298 K. On the basis of these values which of the two substances is apparently the stronger acid?
(b) Determine which of the two substances is the stronger acid in the *absence* of symmetry considerations. [W. F. Bailey and A. S. Monahan, *J. Chem. Educ.* **55**, 489–493 (1978).]

***7.15.** (a) Use the data given in Table 7.7 to determine the standard free energy functions, $-(G_T^0 - H_0^0)/T$, for both the *trans* and *gauche* forms of 1,2-dichloroethane (g) at 500 K. M = 98.95 g/mol. Consider only the frequencies given.
(b) Recognizing that there are two *gauche* forms, P and M, but only one *trans* form, calculate K_P at 500 K for 2*Trans*(g) $\rightleftharpoons$ P-*gauche*(g) + M-*gauche*(g) from the free energy functions in (a), given that the ground-state energy of the *trans* form is 5.0 kJ/mol less than that of the *gauche* forms.
(c) What fraction of the *trans* form converts to *gauche* forms at equilibrium at 500 K?

Chapter Eight

ANSWERS TO PROBLEMS

CHAPTER ONE

1.1. Imagine that box 1 is divided by partitions into n_1 compartments, box 2 into n_2 compartments, etc. There will thus be a total of N compartments altogether, one for each object. There will be N possible locations for the first object; for each of these there will be $N - 1$ locations for the second object; for each of these there will be $N - 2$ locations for the third; and so on. The total number of ways of placing the N objects in the boxes will thus be $N!$ *if* different permutations within any one box count as "ways." But permutations within a given box do *not* count as "ways," and $N!$ will be too great by a factor equal to the number of ways of permuting n_1 objects within box 1, namely, $n_1!$ It will also be too great by a factor equal to the number of ways of permuting n_2 objects within box 2, namely $n_2!$, and so on. Therefore $N!$ will be too great by a factor of $n_1!n_2!n_3!\ldots$ or $\Pi_i n_i!$ Hence $N!$ must be divided by $\Pi_i n_i!$ to give the correct number of ways.

1.2. (*a*) With $N = 8$ and $E = 4$ quanta we have only the possibilities given in Table 8.1.

(b) Macrostate No. (4) is the most probable.

(c) Macrostates (1), (2), and (3) have inverted populations. If these had been eliminated on this basis we would still have been left with the choice of (4) or (5). Of these, however, the one with the more gradual falloff, namely, (4) is the more likely candidate.

Table 8.1. Macrostates for $N = 8$ and $E = 4$

Macrostate	n_0	n_1	n_2	n_3	n_4	W
(1)	7	0	0	0	1	8
(2)	6	1	0	1	0	56
(3)	6	0	2	0	0	28
(4)	5	2	1	0	0	168
(5)	4	4	0	0	0	70

1.3. Taking $N! = (N/2)^N$ still yields $d \ln N! = 0$ in (1.7), since N is a constant. Taking every $n_i! = (n_i/2)^{n_i}$ gives $\ln n_i! = n_i \ln n_i - n_i \ln 2$, instead of $n_i \ln n_i - n_i$ as required by the Stirling approximation. Therefore

$$\begin{aligned} d \sum_i \ln n_i! &= d \sum_i (n_i \ln n_i - n_i \ln 2) \\ &= \sum_i d(n_i \ln n_i) - \ln 2 \sum_i dn_i \\ &= \sum_i (dn_i + \ln n_i dn_i) - \ln 2 \sum_i dn_i \\ &= \sum_i \ln n_i dn_i \qquad \text{since} \sum_i dn_i = 0 \end{aligned}$$

Thus the use of the cruder approximation will give the same result for the Boltzmann distribution law.

1.4. Integration of the Clausius–Clapeyron equation gives $\ln p = -\Delta H^v/RT + \text{const}$. Taking antilogs yields $p = (\text{const})e^{-\Delta H_v}/RT = (\text{const})e^{-\Delta hv/kT}$ if Δh^v is the enthalpy of vaporization per molecule. Replacing ρ by MP/RT (by the ideal gas law) in the barometric relation, M being the molecular weight, gives $d \ln P/dz = -Mg/RT = -mg/kT$ if m is the mass of one molecule. Integration gives $P = (\text{const})e^{-mgz/kT}$. Since Δh^v and mgz are both molecular energies, the right sides of both of the above equations, as well as the Boltzmann distribution law, have the form (const)exp($-$molecular energy/kT).

1.5. (a) Since $q = e^{-x/2} + e^{-3x/2} + e^{-5x/2} + \cdots = e^{-x/2}(e^0 + e^{-x} + e^{-2x} + \cdots)$, and since the factor in parentheses is the same as $(1 - e^{-x})^{-1}$, $q = e^{-x/2}/(1 - e^{-x})$.
(b) As x increases indefinitely $e^0 + e^{-x} + e^{-2x} + \cdots$ approaches unity but $e^{-x/2}$ approaches zero. The lower limit of q (here) is thus zero, not unity. (This is a consequence of the ground-state energy, ϵ_0, being nonzero.

1.6. 0.9996 at 300 K, 0.9016 at 1000 K.

1.7. $g = 6$.

1.8. (a) $q = 2.57$.
(b) The second energy level.
(c) 0.47.

1.9. (a) For System I, $E/N = 3/45 = 0.067$ quanta per particle. For System II $E/N = 2/30 = 0.067$ quanta per particle, or the same as for System I.
(b) The falloff is strictly exponential only for very large N's.
(c) $n_0 = 42 + 28 = 70$, $n_1 = 3 + 2 = 5$, $n_2 = 0 + 0 = 0$. For this $W = (45!/42!3!0!)(30!/28!2!0!) = 6{,}172{,}650$.
(d) $W = 75!/70!5! = 17{,}259{,}390$. Since there *is* a distribution with a greater W than for the one in (c) part ($17{,}259{,}390 > 6{,}172{,}650$) any one with a lower value cannot be the equilibrium one—it must change by an exchange of energies toward the most probable distribution.

1.10. (a) 7.
(b) 118,755.
(c) $n_0 = 20$, $n_1 = 5$, $n_2 = n_3 = n_4 = n_5 = 0$.
(d) 0.447.

CHAPTER TWO

2.1. (a) $W'/W_{eq} = (9/10)^{6.02\times10^{23}\times10^{-10}} = 10^{-2.75\times10^{12}}$ [compare (2.3)].

(b) $W'/W_{eq} = 0.00176$.

2.2. (a) $W_{tot} = 10^{1.16\times10^{19}}$ at 1 K; $W_{tot} = 1$ at 0 K. Rise of 1 K has increased W_{tot} by a factor of $10^{10^{19}}$.

(b) $10^{-3.15\times10^{16}}$.

2.3. (a) $W_{tot} = e^{S/k}$ by (2.7). At 298 K, $W_{tot} = e^{5.69/1.38\times10^{-23}} = 10^{1.79\times10^{23}}$. At 498 K, similarly, $W_{tot} = 10^{3.66\times10^{23}}$. Therefore for the two systems, if still independent, $W_{tot} = W_{298} \times W_{498} = 10^{5.45\times10^{23}}$.

(b) At thermal equilibrium $T = 410$ K and $S = 2 \times 9.03$ J/K since there are two moles at this temperature. W_{tot} now equals $e^{2(9.03)/1.38\times10^{-23}} = 10^{5.69\times10^{23}}$.

(c) W_{tot} has increased from $10^{5.45\times10^{23}}$ to $10^{5.69\times10^{23}}$ as a result of removing the partition separating the two systems, confirming that the process is spontaneous. [This can be shown also, of course, by demonstrating that ΔS for the process is positive: $\Delta S = 2(9.03) - (5.69 + 11.63) = 0.74$ J/K mol.]

2.4. The overall change is $H_2O)(l, 263\text{ K}) \rightarrow 0.875\ H_2O(l, 273\text{ K}) + 0.125\ H_2O(s, 273\text{ K})$ which may be regarded as the sum of

$$H_2O(l, 263\text{ K}) \rightarrow H_2O(l, 273\text{ K}) \quad (1)$$

and

$$0.125H_2O(l, 273\text{ K}) \rightarrow 0.125\ H_2O(s, 273\text{ K}) \quad (2)$$

For (1), $\Delta S = 75.3 \ln (273/263) = 2.81$ J/K. For (2), $\Delta S = 0.125(-6025/273) = -2.76$ J/K. Adding (1) and (2) gives $\Delta S = 0.05$ J/K, which is greater than zero.

2.5. $(1/W_{tot})(\partial W_{tot}/\partial E)_V = 1/kT.$

2.6. (a) Combination of (1.2) and (2.7) gives $S = k \ln [(N + E - 1)!/(N - 1)!E!]$ If one quantum is added E increases by 1 and

$$\begin{aligned}\Delta S &= k \ln\{[(N + E + 1 - 1)!/(N - 1)!(E + 1)!]/\\&\qquad[(N + E - 1)!/(N - 1)!E!]\}\\&= k \ln [(N + E)/(E + 1)] \cong k \ln[(N + E)/E]\end{aligned}$$

(b) Since, by (2.12), $T = (\partial E/\partial S)_V \cong (\Delta E/\Delta S)_V$ we can write

$$T = \frac{\text{energy of one quantum}}{k \ln [(N + E)/E]}$$

CHAPTER THREE

3.1.

$$\begin{array}{r|l} 1 - \mathrm{e}^{-x} \; & \; 1 \qquad\qquad \underline{|\, 1 + \mathrm{e}^{-x} + \mathrm{e}^{-2x} + \mathrm{e}^{-3x} + \cdots} \end{array}$$

$$\begin{array}{l} \underline{1 - \mathrm{e}^{-x}} \\ \quad \mathrm{e}^{-x} \\ \quad \underline{\mathrm{e}^{-x} - \mathrm{e}^{-2x}} \\ \qquad \mathrm{e}^{-2x} \\ \qquad \underline{\mathrm{e}^{-2x} - \mathrm{e}^{-3x}} \\ \qquad\quad \mathrm{e}^{-3x} \\ \qquad\quad \text{etc.} \end{array}$$

3.2. Since $g_0 = 1$, $g_1 = 3$, $g_2 = 6$, $g_3 = 10$, etc.,

$$\begin{aligned}q_v &= e^{-3h\nu/2kT} + 3e^{-5h\nu/2kT} + 6e^{-7h\nu/2kT} + 10e^{-9h\nu/2kT} + \cdots\\&= e^{-3\theta/2T} + 3e^{-5\theta/2T} + 6e^{-7\theta/2T} + 10e^{-9\theta/2T} + \cdots\\&= e^{-3\theta/2T}[1 + 3e^{-\theta/T} + 6e^{-2\theta/T} + 10e^{-3\theta/T} + \cdots]\\&= e^{-3\theta/2T}(1 - e^{-\theta/T})^{-3} = [e^{-\theta/2T}/(1 - e^{-\theta/T})]^3\end{aligned}$$

Therefore, by (3.2),

$$E = NkT^2 \left\{ \frac{\partial}{\partial T} \ln [e^{-\theta/2T}/(1 - e^{-\theta/T})]^3 \right\}_V$$

Replacing N by L (for one mole of solid) and performing the differentiation indicated yields

$$E = 3LkT^2 \{(\theta/2T^2) + [(\theta/T^2)e^{-\theta/T}/(1 - e^{-\theta/T})]\}$$
$$= \tfrac{3}{2}R\theta + [3R\theta e^{-\theta/T}/(1 - e^{-\theta/T})] = \tfrac{3}{2}R\theta + \frac{3R\theta}{e^{\theta/T} - 1}$$

as before.

3.3. With energy zero equal to the $v = 0$ state, $\epsilon_0 = 0$, $\epsilon_1 = h\nu$, $\epsilon_2 = 2h\nu$, etc., so

$$q_v = 1 + e^{-h\nu/kT} + e^{-2h\nu/kT} + e^{-3h\nu/kT} + \ldots$$
$$= 1 + e^{-\theta/T} + e^{-2\theta/T} + e^{-3\theta/T} + \ldots = (1 - e^{-\theta/T})^{-1}$$
$$E = 3NkT^2 \left[\frac{\partial}{\partial T} \ln (1 - e^{-\theta/T})^{-1} \right]_V$$
$$= 3NkT^2(1 - e^{-\theta/T})(-1)(1 - e^{-\theta/T})^{-2}(-e^{-\theta/T})(\theta/T^2)$$
$$= 3Nk\theta(1 - e^{-\theta/T})^{-1}e^{-\theta/T} = 3Nk\theta(e^{\theta/T} - 1)^{-1}$$

which, for one mole, becomes $3R\theta/(e^{\theta/T} - 1)$.

3.4. E is $\tfrac{3}{2}R\theta$ at 0 K and increases indefinitely as T increases. The *slope* of the E vs. T graph equals C_V at the same temperature, which increases from zero at $T = 0$ to a limiting value of $3R$ per mole at $T = \infty$. Furthermore, by Figure 3.1 the slope is always positive and increases slowly with T at first, and then quickly, and then more slowly, eventually becoming constant.

3.5. (a) From Figure 3.1, $T/\theta = 0.43$ (more accurately 0.427) where $C_V = 16.11$ J/K mol, so $\theta = 100/0.427 = 234$ K.

C_V at 298 K, using (3.24) with $\theta/T = 234/298 = 0.785$, is found to be 23.70 J/K mol.

(b) Using (3.26) and $V = 63.5 \times 10^{-3}/8.93 \times 10^3 = 7.11 \times 10^{-6}$ m^3/mol gives $C_P - 23.70 = (4.95 \times 10^{-5})^2(7.11 \times 10^{-6})(298)(1.013 \times 10^5)/(7.50 \times 10^{-7}) = 0.70$ J/K mol, so that $C_P = 24.40$ J/K mol.

(c) This agrees well with the experimental value.

3.6. $C_P = 25.3$ J/K mol.

3.7. (a) Boron, with the larger θ, and therefore the larger oscillator frequency, and therefore the wider spacing of the energy levels, must have stronger interatomic bonds than beryllium.

(b) With increase in T, C_V will rise more rapidly for beryllium than for boron.

3.8. $T = 150$ K.

3.9. $S = 3R\left[\dfrac{\theta/T}{e^{\theta/T} - 1} - \ln\left(1 - e^{-\theta/T}\right)\right].$

3.10. $S = 20.1$ J/K mol.

3.11. (a) Since $\alpha = (1/V)/(\partial V/\partial T)_P = 7.3 \times 10^{-5}$ K^{-1}, and $V = 0.0120/3.51 \times 10^3$ m^3/mol, $(\partial V/\partial T)_P = 7.3 \times 10^{-5} \times 0.0120/3.51 \times 10^3 = 2.50 \times 10^{-10}$ m^3/K mol $= 2.50 \times 10^{-10}$ J Pa^{-1} K^{-1} mol^{-1}. Therefore $\Delta S = -(\partial V/\partial T)_P \Delta P = -2.50 \times 10^{-10}(16{,}000 - 1)(101{,}325) = -0.405$ J/K mol.

(b) The entropy is thus decreased by (0.405/2.439)100 or 16.6%.

3.12. If A is the element with the smaller oscillator frequency it will require smaller quanta of energy than B. The same amount of energy added to both A and B will therefore represent the addition of more quanta to A than to B. More added quanta means more ways in which the added quanta can be distributed, and so W_{tot} will increase more in A than it will in B. Hence ln W_{tot} will increase more in A and so will the entropy. It follows from (2.14) that the average temperature of B is greater than that of A. Since both A and B were initially at the same temperature, B must have undergone the greater increase in temperature and so have the lower C_V.

3.13. (a) For both A and B, $W_{tot} = 1$ and $S = 0$.
(b) For both A and B at 0 K, $n_0 = 100$. For A, after the addition of 13.2×10^{-22} J = 2 quanta, two configurations are possible:

$$n_0 = 99, \quad n_1 = 0, \quad n_2 = 1, \quad W = 100$$
$$n_0 = 98, \quad n_1 = 2, \quad n_2 = 0, \quad W = 4950$$
$$W_{tot} = 5050$$

For B, after the addition of 13.2×10^{-22} J = 1 quantum, only one configuration is possible:

$$n_0 = 99, \quad n_1 = 1, \quad W = 100 = W_{tot}$$

(c) For A, W_{tot} has increased by $5050 - 1 = 5049$; for B, it has increased by $100 - 1 = 99$. Therefore W_{tot} has increased more for A.
(d) Using the result in (c) with the Boltzmann–Planck equation shows that the entropy increased more in A.
(e) For finite changes (2.14) may be written $(\Delta \ln W_{tot}/\Delta E)_V = 1/kT_{av}$, where T_{av} is the average of the initial and final temperatures. Since ΔE is the same for both but $\Delta \ln W_{tot}$ is greater for A, T_{av} must be smaller for A. However, the temperatures of both A and B were the same initially. Hence the final temperature must have been lower in A and the rise in temperature must have been smaller in A.

3.14. **Method 1:** $\theta/T = 770/616 = 1.250$. Substitution of this in the entropy expression given in the answer to Problem 3.9, which comes from (3.12), yields

$$S_{616} = 3R\left[\frac{1.250}{e^{1.250} - 1} - \ln\left(1 - e^{-1.250}\right)\right] = 20.94 \text{ J/K mol}$$

Method 2: W_{tot} for a three-dimensional oscillator is the cube of W_{tot} for a one-dimensional one; that is, $\ln W_{tot(3)} = 3 \ln W_{tot(1)}$. For a mole of one-dimensional oscillators, since $q_v = e^{-\theta/2T}/(1 - e^{-\theta/T})$ and $\theta/T = 1.250$, $q_v = 0.750$. Therefore $n_0 = Le^{-\theta/2T}/q_v = 4.298 \times 10^{23}$, $n_1 = Le^{-3\theta/2T}/q_v = 1.231 \times 10^{23}$, etc. Thus we find the following populations: $n_0 = 4.298 \times 10^{23}$; $n_1 = 1.231 \times 10^{23}$; $n_2 = 3.528 \times 10^{22}$; $n_3 = 1.011 \times 10^{22}$; $n_4 = 2.896 \times 10^{21}$; $n_5 = 8.297 \times 10^{20}$; $n_6 = 2.377 \times 10^{20}$; $n_7 = 6.810 \times 10^{19}$. Since $\ln W_{tot} = \ln L! - \sum n_i \ln n_i!$ for L one-dimensional oscillators, which reduces to $L \ln L - L - \sum n_i (n_i \ln n_i - n_i) = L \ln L - \sum n_i \ln n_i$, $\ln W_{tot}$ becomes $6.0220 \times 10^{23} \ln (6.0220 \times 10^{23}) - [4.298 \times 10^{23} \ln (4.298 \times 10^{23}) + 1.231 \times 10^{23} \ln (1.231 \times 10^{23}) + \cdots] = 5.010 \times 10^{23}$, and so $\ln W_{tot}$ for three-dimensional oscillators $= 3 \times 5.010 \times 10^{23} = 1.5030 \times 10^{24}$. It follows that $S = 1.381 \times 10^{-23} \times 1.5030 \times 10^{24} = 20.76$ J/K mol, which agrees with the result of Method (1) as closely as can be expected.

3.15. $E - E_0 = 5299$ J/mol; $C_V = 23.74$ J/K mol; $S = 33.87$ J/K mol.

3.16. (a) $\theta_D/T = 1860/298.1 = 6.24$, for which $D = 0.0716$ from Table 3.3. Substitution in (3.32) gives $S = 2.43$ J/K mol.

(b) The entropy is unusually small. Because of the strong covalent bonds the energy needed to excite the oscillators is unusually large, so that fewer oscillators can be excited by the input of a given quantity of heat. This reduces C_V and therefore S.

CHAPTER FOUR

4.1. In (4.2) the $(g_i - 1)!$ in the denominator will cancel with $g_i - 1$ of the $g_i + n_i - 1$ factors in the numerator leaving $W_{\text{bosons}} = \Pi_i(g_i + n_i - 1)(g_i + n_i - 2)(g_i + n_i - 3) \ldots (g_i + n_i - n_i)/n_i!$ which will have n_i factors in the numerator. Since $g_i \gg n_i$ the addition of n_i, or of any numbers less than n_i, to it has hardly any effect on it, so each factor in the numerator may be replaced by g_i to give $W_{\text{bosons}} = \Pi_i g_i^{ni}/n_i!$. In (4.3) the $(g_i - n_i)!$ in the denominator will cancel with $g_i - n_i$ of the g_i factors in the numerator leaving $W_{\text{fermions}} = \Pi_i g_i(g_i - 1)(g_i - 2) \ldots (g_i - n_i + 1)/n_i!$, which will have n_i factors in the numerator. Since $g_i \gg n_i$ the subtraction from it of any numbers equal to or less than $n_i - 1$ has hardly any effect on it, so each factor in the numerator may be replaced by g_i to give $W_{\text{fermions}} = \Pi_i g_i^{ni}/n_i!$.

4.2. $S = 154.735$ J/K mol.

4.3. (a) With three possible translational and two possible rotational levels there will be $3 \times 2 = 6$ possible combinations as follows given in Table 8.2.

(b) A value of $\epsilon_{\text{tot}} = 2$ can arise from *more than one* combination of ϵ_t and ϵ_r, as indicated by an asterisk. Thus the degeneracy of this level is $3 + 2 = 5$.

(c) $q_t = 1 + 2e^{-1/kT} + 3e^{-2/kT}$; $q_r = 1 + 2e^{-2/kT}$; $q_t q_r = 1 + 2e^{-1/kT} + 5e^{-2/kT} + 4e^{-3/kT} + 6e^{-4/kT}$, which is the same result as would have been obtained had we known only the ϵ_{tot} values and their degeneracies. *Note:* Degeneracy in ϵ_{tot} arises from two sources: degeneracies in the individual ϵ_t and ϵ_r values and different combinations of ϵ_t and ϵ_r giving the same ϵ_{tot} value.

4.4. Changing V without changing T will not affect E_t for an ideal gas. Hence in (4.38) only the term in q_t will be

Table 8.2. Combinations of Degenerate Energy Levels

ϵ_t	ϵ_r	ϵ_{tot}	g	ϵ_t	ϵ_r	ϵ_{tot}	g
0	0	0	$1 \times 1 = 1$	0	2	2*	$1 \times 2 = 2$
1	0	1	$2 \times 1 = 2$	1	2	3	$2 \times 2 = 4$
2	0	2*	$3 \times 1 = 3$	2	2	4	$3 \times 2 = 6$

affected. Since V is doubled, q_t will be doubled and $\ln(q_t/N)$ will be increased by ln 2. Therefore S_t will be increased by $R \ln 2$ or 5.76 J/K mol and so will S, since q_r and q_v are unaffected by changes in volume.

4.5. $H_t = RT^2 \left(\dfrac{\partial \ln q_t}{\partial T}\right)_V + RT; \qquad H_r = RT^2 \dfrac{d \ln q_r}{dT};$

$$H_v = RT^2 \frac{d \ln q_v}{dT}$$

CHAPTER FIVE

5.1. (a) Using (5.16) gives, for the molar translational entropies of Zn(g) and HCl(g), respectively, 160.9 and 153.6 J/K mol.

(b) The agreement between the calculated and experimental values for Zn is to be expected since it is monatomic and therefore has only translational entropy. The disagreement for HCl reflects the fact that the translational contribution is not the only one to be considered. Since it is diatomic one expects a rotational and possibly also a vibrational contribution to be significant.

5.2. (a) $C = 68.5$ J/K mol.

(b)$S_{t298} = 154.7$ J/K mol, which agrees with the value in Example 5.2.

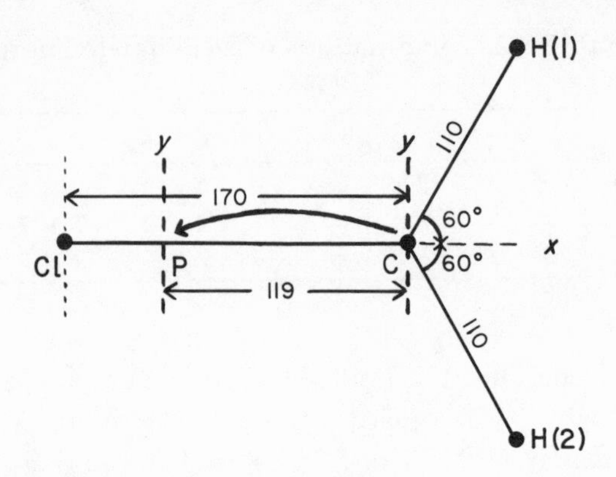

Figure 8.1. The chloromethyl radical

5.3. Because the species is planar and has a twofold axis of symmetry coincident with the C–Cl bond, we know that the center of mass, P, must be on the axis, as in Figure 8.1. Letting the carbon atom be an arbitrary origin with the x axis along the C–Cl bond leads to the following x values for the atoms: Cl $= -170$, C $= 0$, $H_1 = H_2 = 110 \cos 60°$. Using only the first equation of (5.18), since P must be somewhere on the x axis, gives $35.0(-170 - x') + 12.0(0 - x') + 1.0(110 \cos 60° - x') + 1.0(110 \cos 60° - x') = 0$, x' being the coordinate of P on the x axis. The solution of this equation is $x' = -119$ pm. Thus P lies 119 pm to the left of the C atom.

5.4. (a) Since the C–Cl bond is an axis of symmetry it must coincide with a principal axis. As the species is planar a second principal axis must lie in this plane and perpendicular to the plane of the paper, and pass through P. The third principal axis must be perpendicular to both of these axes. In the figure accompanying the answer to Problem 5.3, with the plane of the molecule in the plane of the paper, the x and y axes are as indicated, and the z

axis (not shown) is perpendicular to the plane of the paper.

(b) Since P has been found (in Problem 5.3) to be 119 pm to the left of the carbon atom we may now move the origin to P so that the new coordinate axes coincide with the principal axes. On this basis the coordinates of the atoms now become H(1)(174, 95, 0); H(2)(174, −95, 0); C(119, 0, 0); Cl(−51, 0, 0). [The y coordinates of H(1) and H(2) are given by $\pm 110 \sin 60° = 95$ and -95, respectively.] Using (5.21), (5.22), and (5.23) we then have $I_{xx} = (1.0/L)(95^2 + 0^2) + (1.0/L)[(-95)^2 + 0^2] + (12.0/L)(0^2 + 0^2) + (35.0/L)(0^2 + 0^2) = 3.0 \times 10^{-20}$ g pm^2 $= 3.0 \times 10^{-47}$ kg m^2; $I_{yy} = (1.0/L)(174^2 + 0^2) + (1.0/L)(174^2 + 0^2) + (12.0/L)(119^2 + 0^2) + (35.0/L)[(-51)^2 + 0^2] = 5.3 \times 10^{-19}$ g pm^2 $= 5.3 \times 10^{-46}$ kg m^2; $I_{zz} = (1.0/L)(174^2 + 95^2) + (1.0/L)[174^2 + (-95)^2] + (12.0/L)(119^2 + 0^2) + (35.0/L)[(-51)^2 + 0^2] = 5.6 \times 10^{-19}$ g pm^2 $= 5.6 \times 10^{-46}$ kg m^2. Since these are moments about the principal axes they are also the principal moments of inertia. Therefore $I_A I_B I_C = I_{xx} I_{yy} I_{zz} = 9 \times 10^{-138}$ kg^3 m^6.

(c) $I_{xy} = (1.0/L)(174)(95) + (1.0/L)(174)(-95) + (12.0/L)(119)(0) + (35.0/L)(-51)(0) = 0$; $I_{xz} = (1.0/L)(174)(0) + (1.0/L)(174)(0) + (12.0/L)(119)(0) + (35.0/L)(-51)(0) = 0$; $I_{yz} = (1.0/L)(95)(0) + (1.0/L)(-95)(0) + (12.0/L)(0)(0) + (35.0/L)(0)(0) = 0$.

(d) $I_A + I_B = 3.0 \times 10^{-47} + 5.3 \times 10^{-46} = 5.6 \times 10^{-46} = I_C$ as required.

5.5. $I = 1.380 \times 10^{-45}$ kg m^2.

5.6. (a) This problem is similar to Problem 5.3 but HOD has less symmetry than CH_2Cl. There is now only one plane of symmetry and no axis of symmetry to assist in locating the center of mass, P. The latter will not lie on the bisector of the bond angle, but between it and the D atom and in the plane of the molecule. If, as in Figure

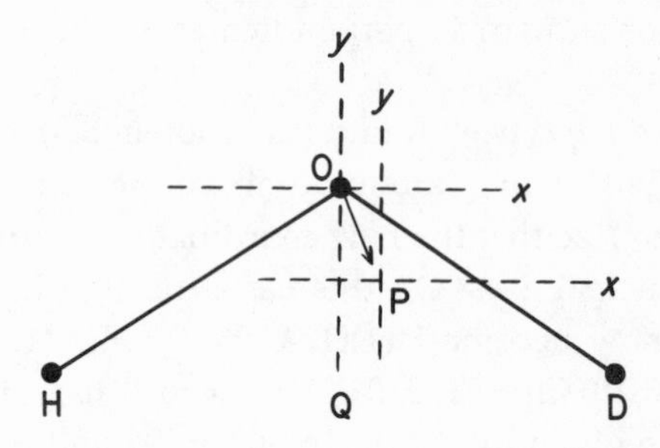

Figure 8.2. The HOD molecule.

8.2, we arbitrarily make the oxygen atom the origin, with the y axis coinciding with the angular bisector, the x axis at right angles to it (with both axes in the plane of the paper), and the z axis perpendicular to both, the atomic coordinates become H(-75.69, -58.58, 0); D(75.69, -58.58, 0); O(0, 0, 0) after noting that HQ = QD = 95.71 sin(104.53/2) = 75.69 pm and that QO = 95.71 cos(104.53/2) = 58.58 pm. From (5.18), with mass in amu, $1.0078(-75.69 - x') + 2.0141(75.69 - x') + 15.9949(0 - x') = 0$ or $x' = 4.01$ pm. Likewise $1.0078(-58.58 - y') + 2.0141(-58.58 - y') + 15.9949(0 - y') = 0$ or $y' = -9.31$ pm.

(b) Since $I_A I_B I_C$ is required, and not I_A, I_B, or I_C individually, we may use (5.31). Unlike Problem 5.4 we have not determined the *orientation* of the principal axes about P, but that is not needed. We transfer the origin from the oxygen atom to P without rotating the coordinate system, giving the following atomic coordinates: H(-79.70, -49.27, 0); D(71.68, -49.27, 0); O(-4.01, 9.31, 0). Application of Eqs. (5.21) to (5.26) gives I_{xx}, I_{yy}, and I_{zz} = 1.448, 2,824, and 4.272×10^{-47} kg m^2, respectively, and I_{xy}, I_{xz}, and I_{yz} = -6.232×10^{-48}, 0, and 0 kg m^2, respectively. (Notice that since *all* the products of inertia are not zero, the axes chosen are not all principal axes, although we know that the chosen z axis is a principal one.) Placing the results in (5.31) yields

$$I_A I_B I_C = \begin{vmatrix} 1.448 \times 10^{-47} & -6.232 \times 10^{-48} & 0 \\ -6.232 \times 10^{-48} & 2.824 \times 10^{-47} & 0 \\ 0 & 0 & 4.272 \times 10^{-47} \end{vmatrix}$$

$$= 1.581 \times 10^{-140} \text{ kg}^3\text{m}^6$$

5.7. $I_A I_B I_C = 2.852 \times 10^{-134}$ kg^3 m^6.

5.8. (a) From (5.35) $I = h^2/8\pi^2\theta_r k = 6.69 \times 10^{-46}$ kg m^2.

(b) From (5.16) $S_t^0 = 155.94$ and, from (5.46), $S_r = 59.91$ J/K mol.

(c) $S_t^0 + S_r = 215.85$ J/K mol, somewhat less than the literature value. Apparently the translational and rotational contributions are not the only ones to be considered. There is likely a contribution from the vibrational energy (compare Problem 5.13).

5.9. Since S_t^0 depends only on M and T the two gases will have nearly the same translational entropy. However, σ is 1 for CO but 2 for N_2, so their rotational entropies will differ by $R \ln 2$ or 5.8 J/K mol, that of CO being the greater.

5.10. (a)

$$q_r = \tfrac{1}{2}[1 + 3e^{-2\theta_r/T} + 5e^{-6\theta_r/T} + 7e^{-12\theta_r/T} + 9e^{-20\theta_r/T} + 11e^{-30\theta_r/T} + 13e^{-42\theta_r/T}].$$

$$E_r = RT^2 \frac{d}{dT} \ln q_r = (RT^2/q_r)(dq_r/dT)$$

$$= \frac{R\theta_r}{2q_r}[3(2)e^{-2\theta_r/T} + 5(6)e^{-6\theta_r/T} + 7(12)e^{-12\theta_r/T} + 9(20)e^{-20\theta_r/T} + 11(30)e^{-30\theta_r/T} + 13(42)e^{-42\theta_r/T}]$$

(b) Since $\theta_r/T = 87.5/298.15 = 0.2935$, q_r (using the above expression) = 1.881 and $E_r = 2230$ J/mol. The classical value is $298.15R = 2478$ J/mol, a difference of about 10%.

(c) From (5.16) $S_t^0 = 117.487$ J/K mol. From (5.39) q_r = 1.881 and from (4.39) $S_r = R \ln 1.881 + (2230/298.15)$ = 12.73, so $S_t^0 + S_r = 130.22$ J/K mol, in good agreement with the literature value, implying little or no contribution from other than translational and rotational sources at this temperature.

(d) From (5.38) $q_r = 298.15/2(87.5) = 1.704$, and $S_r = R \ln 1.704 + R = 12.74$ J/K mol, so $S_t^0 + S_r = 130.23$ J/K mol. Thus, by using (5.38) instead of (5.39), an error of about 10% is introduced into E_r but practically no error into S_r. This is because of a cancellation of errors: in (4.39) q_r is too small but E_r is too large.

5.11. (a) $S_t^0 = 150.142$ J/K mol; $S_r = 56.476$ J/K mol.

(b) S_t^0 and S_r both smaller than for HOD.

(c) $S_{373}^0 = 206.618$ J/K mol; $C_{P373} = 33.26$ J/K mol.

5.12. As σ increases there are fewer distinguishable orientations of the molecule as it rotates, and therefore fewer microstates available. This reduces W and hence S through the Boltzmann–Planck equation.

5.13. $C_V = 30.03$ J/K mol.

5.14. (a) The Einstein model for vibration in solids is essentially the same as that used in Chapter 5, except that in the former a given vibration is effectively the equivalent of three otherwise identical oscillations. Hence the effective N is three times as large. The Einstein model treats distinguishable particles. Although our model of gases assumes the molecules to be indistinguishable, the indistinguishability is incorporated into the translational

contribution leaving the rotational and vibrational treatments effectively those of distinguishable molecules.
(b) The θ's for atomic solids are smaller because the frequencies of vibration are less—due to weaker interatomic forces.

5.15. $H - H_0 = 9710$ J/mol.

5.16. $(H^0_{298} - H^0_0)/298.15 = 38.27$ J/K mol; $-(G^0_{298} - H^0_0)/298.15 = 199.11$ J/K mol; $C^0_{P298} = 55.58$ J/K mol; $S^0_{298} = 237.38$ J/K mol.

CHAPTER SIX

6.1. Since Na is monatomic there are no rotational or vibrational contributions. Equation (5.16) gives $S^0_t = 147.84$ J/K mol, so the remainder must be an electronic contribution. Sodium has one unpaired electron so the ground state must be a doublet ($g_0 = 0$). By (6.2) $S_e = R \ln 2 = 5.76$ J/K mol which, added to S^0_t, gives $S^0_{298} = 153.60$ J/K mol, in agreement with experiment.

6.2. (a) $q_e = 3.115$
(b) $E_e = 518$ J/mol; $S_e = 11.184$ J/K mol.

6.3. Because of the great similarity of H and D atoms there are four nearly identical frozen-in orientations of every CH_3D molecule in the crystalline solid at 0 K. If all four are randomly mixed the resulting entropy will be $k \ln 4^L = R \ln 4 = 11.5$ J/K mol—close to the observed residual entropy of 11.6.

6.4. $X_{para} = 0.499$.

6.5. (a) q_r (para), calculated as in Problem 6.4, but using $\theta_r/T = 0.5833$, is 1.151.

$$E_r = RT^2 \frac{d \ln q_r}{dT} = \frac{R\theta_r}{q_r}(30e^{-6\theta_r/T} + 180e^{-20\theta_r/T} + \cdots)$$

so

$$C_r = \frac{dE_r}{dT} = R\theta_r \frac{d}{dT}[(30e^{-6\theta_r/T} + 180^{-20\theta_r/T} + \cdots)/q_r] = 11.68 \text{ J/K mol}$$

(b) q_r (ortho), calculated analogously to para in part (a) is 2.822 if nuclear degeneracy be included.

$$C_r = R\theta_r \frac{d}{dT}[(6e^{-2\theta_r/T} + 84e^{-12\theta_r/T} + \ldots)/q_r] = 1.91$$

Therefore C_r for the 1:3 mixture = 0.25(11.68) + 0.75(1.91) = 4.35 J/K mol.

6.6. (a) $q_r = 6 \sum_{J \text{ even}} (2J + 1)e^{-J(J+1)\theta_r/T} + 3 \sum_{J \text{ odd}} (2J + 1)e^{-J(J+1)\theta_r/T}$

(b) (i) At 0 K $J = 0$ so all the D_2 is ortho and $X_{\text{ortho}} = 1$. (ii) At high enough temperatures $X_{\text{ortho}} = 6/(6 + 3) = 0.67$.

6.7. $S^0_{298} = 358.89$ J/K mol, slightly larger than the experimental value, possibly because the methyl groups are not rotating freely.

6.8. The complete symmetry numbers are 18 for (i), 162 for (ii), 2 for (iii), and 972 for (iv).

6.9. (a) By (5.11), (5.53), and (6.11), C_{Vt}, C_r, and C_{ir} are $\frac{3}{2}R$, $\frac{3}{2}R$, and $\frac{1}{2}R$, respectively, the sum of which is $\frac{7}{2}R$ To find C_v the contributions of the various vibrational modes are summed using (5.76): For the first mode listed, $\theta_v/T = hc(47{,}600)/k(373.15) = 1.835$, so $C_v/R = (1.835)^2e^{1.835}/$

$(e^{1.835} - 1)^2 = 0.761$. For the succeeding modes $C_v/R =$ 0.653, 0.610, 0.384, 0.268, 0.237, 0.138, 0.129, 0.118, 0.107, 0.084, 0.001, 0.001, 0.001. The sum of the terms for all the modes is 3.492, hence $C_v = 3.492R$. Therefore $C_V = \frac{7}{2}R + 3.492R$, and $C_P = C_V + R = 66.45$ J/K mol, which nearly agrees with the measured value within experimental error and indicates little or no restriction to the internal rotation.

(b) $I_A I_B I_C$ will be needed for computing S_r. We will use the method related to (5.32) in which it is unnecessary to locate the center of mass. Referring to Figure 8.3, let the origin be placed at the C atom, with H_1 in the xz plane, the C–N axis coincident with the x axis, and the O atoms in the xy plane. The atomic coordinates then become: $H_1(36, 0, -103)$; $H_2(36, 89, 51)$; $H_3(36, -89, 51)$; C(0, 0, 0); N(−146, 0, 0); $O_1(-200, 108, 0)$; $O_2(-200, -108, 0)$. It is then found that $A = 404{,}901$, $B = 459{,}707$, $C = 832{,}986$ amu pm^2, $D = E = F = 0$, so $I_A I_B I_C = ABC = 1.550 \times 10^{17}$ amu^3 pm^6 $= 7.10 \times 10^{-136}$ kg m^2. Fur-

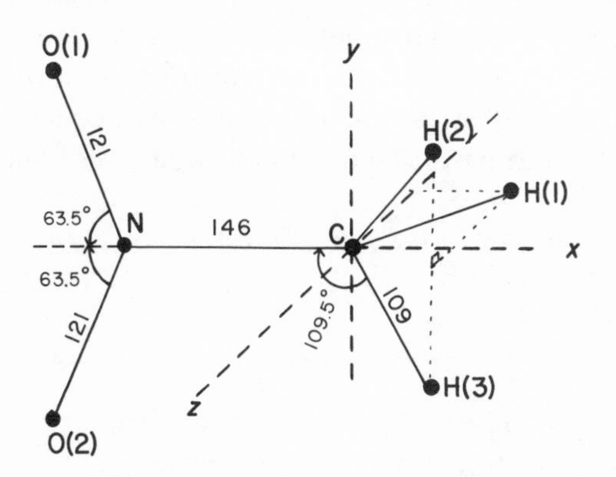

Figure 8.3. The nitromethane molecule.

thermore, the moment of inertia of the CH_3 about the C–N axis (I_1) is $3(1.0/L)(103^2) = 5.3 \times 10^{-20}$ g pm^2 = 5.3×10^{-47} kg m^2, and the moment of inertia of the rest of the molecule about the same axis is $2(16.0/L)(108^2) = 6.2 \times 10^{-19}$ g pm^2 = 6.2×10^{-46} kg m^2. This gives $I_{ir} = 4.9 \times 10^{-47}$ kg m^2.

From (5.16) $S_t^0 = 160.02$ J/K mol. Proceeding exactly as in the answer to Problem 6.7 above, using the above values for $I_A I_B I_C$ and I_{ir}, and remembering that there is only one rotating group, we find $q_r'(q_{ir}') = 3.21 \times 10^5$. With $\sigma_{csn} = 6$ (the same as for toluene, for the same reasons), $S_r + S_{ir} = R \ln (3.21 \times 10^5) - R \ln 6 + \frac{3}{2}R + \frac{1}{2}R = 107.15$ J/K mol. To find S_v the contributions of the various vibrational modes are summed according to (5.80). For example, for the first mode listed, $\theta_v/T = hc(47600)/k(298.15) = 2.297$, so S_v/R for that mode is 0.3628. For the succeeding modes S_v/R = 0.2270, 0.1890, 0.0646, 0.0318, 0.0253, 0.0097, 0.0086, 0.0073, 0.0062, 0.0042, 0, 0, 0. The sum of these terms for all the modes is 0.9365, so $S_v = 0.9365R = 7.786$ J/K mol. Therefore $S_{298}^0 = 160.02 + 7.79 + 107.15 = 274.96$ J/K mol, which is in excellent agreement with the third law result and leads one to believe that any restriction on the internal rotation must be small. The same conclusion was reached in part (a) referring, however, to a temperature 75 K higher.

6.10. The values of $-(G_{298}^0 - H_0^0)/298.15$ for H_2(g), I_2(g), and HI(g) are 101.95, 226.60, and 177.18 J/K mol, respectively.

6.11. (a) $\Delta H_0^0 = -4.55$ kJ/mol HI(g).

6.12. For CH_3D $\sigma = 3$: the number of equivalent positions in space resulting from overall rotation of the molecule. (The three equivalent positions can be obtained by rotating the molecule one revolution about the C–D axis.) On the other hand, rotation about any axis is quenched at 0 K so σ is irrelevant. However, four distinguishable orientations can be frozen in randomly as CH_3D is cooled to 0 K, depending on the four possible orientations of the C–D bond. Therefore $W_{tot} = 4$ for each molecule at that temperature.

CHAPTER SEVEN

7.1. (a) For C_2H_4, Eq. (6.16) gives $[-(G_T^0 - H_0^0)/T]_t = \tfrac{3}{2}R \ln 28.05 + \tfrac{5}{2}R \ln 298.15 - 30.471 = 129.54$ J/K mol; Eq. (6.19) gives

$$[-(G_T^0 - H_0^0)/T]_r = \tfrac{3}{2}R \ln 298.15 + R \ln \frac{(5.7 \times 27.5 \times 33.2 \times 10^{-141})^{1/2}}{4} + 1308.37 = 53.78 \text{ J/K mol}$$

Eq. (6.20) gives

$$[-(G_T^0 - H_0^0)/T]_v = -R\left\{\ln\left[1 - \exp\left(-\frac{0.014388 \times 900 \times 100}{298.15}\right)\right] + \ln\left[1 - \exp\left(-\frac{0.014388 \times 940 \times 100}{298.15}\right)\right] + \cdots\right\} = 0.434$$

after summing eight terms, the only ones out of all twelve modes which contribute appreciably. The factor common to all the exponents, 0.014388, equals hc/k, with h in J s, c in m/s, and k in J/K. The other common factor,

100, converts cm^{-1} to m^{-1}. The sum of all three free energy functions is 183.75 J/K mol.

For C_2H_6, the same calculations give, for the translational and rotational free energy functions, 130.41 and 55.41 J/K mol, respectively. In (6.19), $\sigma = 6$ for C_2H_6. Of the eighteen vibrational modes, one is to be treated as an internal rotation. Of the remaining seventeen only the eleven listed contribute appreciably. The latter give a vibrational free energy function of 0.504 by the same procedure used above for C_2H_4. Using (6.9), with $\sigma_{ir} = 3$, gives $q_{ir} = 2.61$ and an additional contribution to the free energy function of $R \ln 2.61 = 7.976$ J/K mol. The total free energy function for C_2H_6 is thus 130.41 + 55.41 + 0.504 + 7.98 = 194.30 J/K mol.

(b) The calculated and tabulated values for C_2H_4 are close enough to be able to attribute the difference to experimental errors and approximations in the theoretical treatment. For C_2H_6, however, the difference, about 4.9 J/K mol, indicates that the internal rotation is restricted, not free, as a result of a potential energy barrier.

(c) $\Delta G^0_{298} = -129.87 - (189.41 - 184.01 - 102.17)(298.15/1000) = -101.02$ kJ by (6.21). Using (7.1), $-101{,}020 = -8.314(298.15) \ln K_p$, so $K_p = 5.00 \times 10^{17}$.

7.2. $K_p = 1.453 \times 10^3$.

7.3. In (7.17) the translational factor is, according to (7.18), $30.0^3/(28.0 \times 32.0)^{3/2} = 1.0067$. The rotational factor reduces to $(\sigma\theta_r)_{N_2}(\sigma\theta_r)_{O_2}/(\sigma\theta_r)^2_{NO} = (2 \times 2.89 \times 2 \times 2.08)/(1 \times 2.45)^2 = 4.01$ through (5.38). For N_2, through (5.68), $q_v = (1 - e^{-3353/2000})^{-1} = 1.230$; for O_2, similarly, $q_v = 1.485$ and, for NO, $q_v = 1.350$. The vibrational factor is therefore $1.350^2/1.230 \times 1.485 = 0.9978$. For N_2, $q_e = 1.000$ (no degeneracy in the ground state and no low-lying electronic states). For O_2, by (6.1), $q_e = 3e^{-0} + 2\exp(-1.573\times10^{-19}/2000k) = 3.007$; for NO, $q_e = 2e^{-0} + 2\exp(-2.384\times10^{-21}/2000k) = 3.835$. Thus the electronic factor is $3.835^2/1.000 \times 3.007 = 4.891$. Since $a + b = m$ the L factor is unity. Finally, $\Delta H_0^0 = [941.2 + 490.1 - 2(626.1)]\,1000 = 179.1 \times 10^3$ J. Therefore

$$\begin{aligned} K_P &= 1.0067 \times 4.01 \times 0.9978 \\ &\quad \times 4.891\exp(-179.1\times10^3/8.314\times2000) \\ &= 4.14 \times 10^{-4} \end{aligned}$$

7.4. $K_P = 0.148$.

7.5. $K_P = 3.291 \times 10^{-7}T^{5/2} \exp(-1.160\times10^4 I/T)$, with I in

7.6. Rewrite (7.17) as follows:

$$\begin{aligned} K_P = F_t F_r \cdots & \frac{(1 - e^{-\theta_M/T})^{-1}(1 - e^{-\theta_N/T})^{-1} \cdots}{(1 - e^{-\theta_A/T})^{-1}(1 - e^{-\theta_B/T})^{-1} \cdots} \\ & \times \exp[-(D_A+D_B+ \cdots -D_M-D_N- \cdots)L/RT] \end{aligned}$$

where the θ's are the characteristic vibration temperatures, the F's the translational and rotational factors, and the D's are the chemical dissociation energies, D_0, per molecule. Since each $D = D' - \frac{1}{2}h\nu$, where D' is the depth of the potential energy well or spectroscopic dissociation energy (see Figure 5.4), the last factor in the K_P expression can be written

$$\begin{aligned} & \exp[-(1/2)h\nu_M/kT] \exp[-(1/2)h\nu_N/kT] \\ & \cdots \exp[(1/2)h\nu_A/kT] \exp[(1/2)h\nu_B/kT] \\ & \cdots \exp[-(D'_A+D'_B+ \cdots -D'_M-D'_N- \ldots)L/RT] \end{aligned}$$

remembering that $R/L = k$. This can be simplified somewhat to yield

$$\begin{aligned} \frac{e^{-\theta M/2RT}e^{-\theta N/2RT} \cdots}{e^{-\theta A/2RT}e^{-\theta B/2RT} \cdots} & \exp[-(D'_A+D'_B \\ & + \cdots -D'_M-D'_N- \cdots)L/RT] \end{aligned}$$

Replacing the last factor in the original expression by this result, rearranging, and recognizing that $(D'_A + D'_B \cdots - D'_M - D'_N - \cdots)L$ is the same as ΔH_0^0 when the vibrationless molecules are the energy zeros, gives

$$K_P = F_t F_r \frac{e^{-\theta_M/2T}(1 - e^{-\theta_M/T})^{-1}\, e^{-\theta_N/2T}(1 - e^{-\theta_N/T})^{-1} \cdots}{e^{-\theta_A/2T}(1 - e^{-\theta_A/T})^{-1}\, e^{-\theta_B/2T}(1 - e^{-\theta_B/T})^{-1} \cdots} e^{-\Delta H_0^0/RT}$$

But each pair of factors, $e^{-\theta_i/2T}(1 - e^{-\theta_i/T})^{-1}$, is the q_v for the given species on the basis of the new energy zero, as in (5.69). Hence

$$K_P = F_t F_r \cdots \frac{q_{vM} q_{vN} \cdots}{q_{vA} q_{vB} \cdots} e^{-\Delta H_0^0/RT}$$

which is the same as the original expression but with altered meanings for q_v and ΔH_0^0.

7.7. The species on the right is a more rigid molecule than that on the left and would therefore be expected to have fewer accessible energy levels and therefore the smaller entropy. At high enough temperatures the entropy considerations will outweigh the energy considerations, and the equilibrium will be expected to lie to the left.

7.8. Applied to the ground states of A and B, (1.17) becomes

$$n_{0A} = (N_A/q_A^0)\, e^{-\epsilon_{0A}/kT} \quad \text{and} \quad n_{0B} = (N_B/q_B^0)\, e^{-\epsilon_{0B}/kT}$$

in the standard state. If the respective ground states are the energy zeros these equations reduce to $n_{0A} = N_A/q_A^0$ and $n_{0B} = N_B/q_B^0$. Then $K_P = N_B/N_A = n_{0B}q_B^0/n_{0A}q_A^0$. But by the Boltzmann distribution

$$\frac{n_{0B}}{n_{0A}} = \frac{e^{-\epsilon_{0B}/kT}}{e^{-\epsilon_{0A}/kT}} = e^{-(\epsilon_{0B}-\epsilon_{0A})/kT} = e^{-\Delta\epsilon_0/kT} = e^{-\Delta\epsilon_0 L/RT}$$
$$= e^{-\Delta H_0^0/RT}$$

Substituting this result into the expression for K_P given above yields

$$K_P = \frac{q_B^0}{q_A^0} e^{-\Delta H_0^0/RT}$$

7.9. $K_P = 3.99$.

7.10. (a) In (7.17) the translational factor is $19.02^3/18.01^{3/2} \times 20.03^{3/2} = 1.0042$. The rotational factor will, through (5.48), be

$$\frac{(I_A I_B I_C)_{\text{HOD}}}{(I_A I_B I_C)^{1/2}_{\text{H2O}} (I_A I_B I_C)^{1/2}_{\text{D2O}}} \times \frac{\sigma_{\text{H2O}} \sigma_{\text{D2O}}}{\sigma^2_{\text{HOD}}}$$
$$= \frac{1.21 \times 3.06 \times 4.27}{(1.02 \times 1.92 \times 2.94)^{1/2} (1.84 \times 3.83 \times 5.67)^{1/2}} \times \frac{2 \times 2}{1}$$
$$= 4.17$$

The vibrational factor will be 1.000 if the ground vibrational states are taken as the energy zeros, since vibrational contributions are being neglected. (This assumption introduces some error at 800 K.) Since the three species differ only isotopically the potential energy curve is the same for them all. ΔH_0^0 can therefore be found as the difference between the zero-point energies, namely, 2(201.79 − 231.92 − 169.58 = 2.080 kJ, giving exp $(-2080/8.314 \times 800) = 0.731$. Hence $K_P = 1.0042 \times 4.17 \times 1.000 \times 0.731 = 3.06$.

(b) The value of K_P found in Example 7.5 differs from this by about 30% but it is still a fair estimate. The principal cause of the discrepancy is the erroneous assumption made in Example 7.5 that $\Delta H_0^0 = 0$.

7.11. There are the translational contributions of all three species and the rotational contribution of the KBr to consider. For KBr, by (7.11), $q_t^0/L = 2.5607 \times 10^{-2}$

$(119)^{3/2}(700)^{5/2} = 4.31 \times 10^8$. Similarly $q_t^0/L = 8.08 \times 10^7$ for K^+ and 2.37×10^8 for Br^-. The moment of inertia for KBr is, by (5.27), $\{(0.039/L)(0.080/L)[(0.039 + 0.080)/L]\}$ $(294 \times 10^{-12})^2 = 3.76 \times 10^{-45}$ kg m^2, so that $q_r = 1.738 \times 10^{47} \times 3.76 \times 10^{-45}/1 = 6.53 \times 10^3$, according to (7.13). Therefore q^0/L for KBr $= (q_t^0/L)q_r = 4.31 \times 10^8 \times 6.53 \times 10^3 = 2.817 \times 10^{12}$. $\Delta H_0^0 = \exp(-472{,}000/8{,}314 \times 700)$ 5.99×10^{-36}. Insertion of these results in (7.8) yields

$$K_P = \frac{8.08 \times 10^7 \times 2.37 \times 10^8}{2.817 \times 10^{12}} \times 4.99 \times 10^{-36} = 4.07 \times 10^{-32}$$

7.12. (a) $\Delta H_0^0 = 120.1$ kJ.
(b) 1.24 eV.

7.13. (a) $I_A I_B I_C = 5.5 \times 10^{-140}$ kg m^2.
(b) 159.0 J/K mol.
(c) 1.5×10^{-139} kg m^2; 152.5 J/K mol.
(d) 85.

7.14. (a) On the basis of their K_a values the stronger acid is the one with the larger dissociation constant, namely, 2,4-dinitrophenol.
(b) Assuming that an expression for K_a analogous to (7.17) is valid for both substances, the symmetry numbers in the rotational factor can be isolated as the factor

$$\frac{\sigma_{\text{undissociated acid}}}{\sigma_{\text{anion}}\sigma_{\text{H+}}}$$

For benzoic acid $\sigma_{\text{csn}} = 2$, for the benzoate ion it is 4, and for H^+ it is 1. For 2,4-dinitrophenol $\sigma_{\text{csn}} = 4$, for the 2,4-dinitrophenolate ion it is 4, and for H^+ it is 1. Thus the symmetry factors for benzoic acid and 2,4-dinitrophenol are $2/4 \times 1 = 1/2$ and $4/4 \times 1 = 1$, respectively. Dividing the K_a for benzoic acid by that for 2,4-dinitrophenol gives $6.31 \times 10^{-5}/8.13 \times 10^{-5} = 0.776$, which must equal the ratio excluding symmetry multiplied by the ratio of the symmetry factors, namely $\frac{1}{2}/1 = \frac{1}{2}$. Therefore

$$\frac{K_a \text{ for benzoic acid, excluding symmetry factor}}{K_a \text{ for 2,4-dinitrophenol, excluding symmetry factor}} = 2 \times 0.776 = 1.55$$

As this is greater than unity, benzoic acid would be the stronger acid of the two were it not for symmetry considerations.

7.15. (a) *Trans:* 266.44 J/K mol; *gauche:* 265.59 J/K mol.
(b) 0.07.
(c) 0.35.

APPENDIX

PHYSICAL CONSTANTS[1]

Avogadro number	L	$= 6.022045 \times 10^{23}\ \text{mol}^{-1}$
Boltzmann constant	k	$= 1.380662 \times 10^{-23}\ \text{J K}^{-1}$
Gas constant	R	$= 8.31441\ \text{J K}^{-1}\ \text{mol}^{-1}$
		$= 0.082057$ liter atm/K mol
Planck constant	h	$= 6.626176 \times 10^{-34}\ \text{J}\cdot\text{s}$
Velocity of light	c	$= 2.99792458 \times 10^{8}\ \text{m}\cdot\text{s}^{-1}$

CONVERSION FACTORS

1 calorie = 4.1840 J

1 eV = 1.6020×10^{-19} J

1 atm = 101,325 Pa = 101,325 N m^{-2}

1 J = 10^{7} erg

1 liter atm = 101.325 J

$hc/k = 0.0143879\ \text{m}\cdot\text{K}$

[1]E. R. Cohen and B. N. Taylor, The 1973 Least-Squares Adjustment of the Fundamental Constants, *J. Phys. Chem. Ref. Data* **2,** 663–734 (1973).

INDEX